The BEACHCOMBER'S GUIDE to SEASHORE LIFE of CALIFORNIA

To my parents, Albert and Liz

The BEACHCOMBER'S GUIDE to SEASHORE LIFE of CALIFORNIA

J. DUANE SEPT

HARBOUR PUBLISHING

Copyright © 2002 J. Duane Sept

All rights reserved. No part of this publication may be reproduced, stored in a retrieval system or transmitted, in any form or by any means, without prior permission of the publisher or, in case of photocopying or other reprographic copying, a licence from CANCOPY (Canadian Reprography Collective), 214 King Street West, Toronto, Ontario M5H 3S6.

Published by
HARBOUR PUBLISHING
P.O. Box 219
Madeira Park, BC Canada
V0N 2H0

Cover photos and all other photos in this book copyright Duane Sept, except for the photograph of the California grunion (p. 217), copyright Mike Brock

Cover design, page design and composition by Martin Nichols

Printed in Hong Kong by Prolong Press Limited

The Canada Council for the Arts since 1957 | Le Conseil des Arts du Canada depuis 1957

Harbour Publishing acknowledges the financial support of the Government of Canada through the Book Publishing Industry Development Program (BPIDP), the Canada Council for the Arts, and the Province of British Columbia through the British Columbia Arts Council, for its publishing activities.

National Library of Canada Cataloguing in Publication Data

Sept, J. Duane, 1950–
  Beachcomber's guide to seashore life of California

  Includes index.
  ISBN 1-55017-251-4

  1. Seashore biology—California—Identification. I. Title.
QH95.7.S42 2002   578.769'9'09794   C2001-911670-5

Contents

Introduction

Understanding Tides... 8
Understanding Intertidal Habitats 10
Harvesting Shellfish 13
Protecting Our Marine Resource 13
Observing Intertidal Life...................................... 14
A Note of Caution 14
Getting the Most Out of This Guide 15

Field Guide

Sponges *(Phylum Porifera)*............ 16

Hydroids, Jellies, Sea Anemones, Comb Jellies
(Phyla Cnidaria and Ctenophora)........ 20

    Hydroids................................... 21
    Large Bell-shaped Jellies................... 24
    Comb Jellies 27
    Sea Anemones and Cup Corals 27

Marine Worms
(Phyla Platyhelminthes, Nemertea, Annelida, Sipuncula and Echiura)....... 36

    Flatworms................................. 37
    Ribbon Worms 39
    Segmented Worms 43
    Peanut Worms 55
    Echiuran Worms........................... 56

continued...

Mollusks and Lampshells
(Phyla Mollusca & Brachiopoda) 57

    Chitons 58
    Gastropods 71
    Nudibranchs and Allies 112
    Bivalves 125
    Octopods and Squids 159
    Lampshells 160

Arthropods *(Phylum Arthropoda)* 161

    Barnacles 162
    Shrimps, Crabs and Allies 166

Moss Animals *(Phylum Bryozoa)* 188

Spiny-Skinned Animals
(Phylum Echinodermata) 192

    Sea Stars 193
    Brittle Stars 198
    Urchins and Sand Dollars 202
    Sea Cucumbers 204

Sea Squirts *(Phylum Urochordata)* ... 207

Fishes *(Phylum Chordata)* 214

continued . . .

Seaweeds
(Phyla Chlorophyta, Phaeophyta, Rhodophyta) 221

    Green Algae 222
    Brown Algae 226
    Red Algae 237

Flowering Plants
(Phylum Anthophyta) 251

Best Beachcombing Sites in California 255
  Key to Site Descriptions 255
  North Coast 256
    Del Norte County 257
    Humboldt County 258
    Mendocino County 259
    Sonoma County 261
    Marin County 263
  Central Coast 265
    San Mateo County 266
    Santa Cruz County 267
    Monterey County 268
    San Luis Obispo County 272
    Santa Barbara County 274
    Ventura County 276
  South Coast 277
    Los Angeles County 278
    Orange County 282
    San Diego County 284

Acknowledgements 288

Further Reading 289

Index 290

Introduction

The California coast is one of the world's richest, most diverse habitats for intertidal marine life. Hundreds of species and subspecies of animals and plants live along these shores, and each of them has developed a unique niche in which it lives, coexisting with its neighbors. To learn what these species are and how they are interrelated is a step toward learning how all the parts of the world work together in the giant puzzle we call life.

The intertidal zone—that part of the shoreline that is submerged in water at high tide and exposed at low tide—is a particularly gratifying place to observe both wildlife and plant life. Species are diverse, abundant and endlessly interesting, and many of them can be observed easily without any special knowledge or equipment. Some are animals that are found both intertidally and subtidally but whose appearance is completely transformed out of water. Anemones, for instance (see pp. 27–35), are often seen on the beach with their tentacles closed, and some marine worms (pp. 36–56) close their tentacles or leave distinctive signs on a beach when the tide recedes. Other species, such as the moonglow anemone (p. 28) or Merten's chiton (p. 61), occur in several color forms.

This guide is designed to enhance your experience of observing and identifying animal and plant species in the many fascinating intertidal sites of California. Many of these areas are so rugged they seem indestructible, but in fact they are fragile ecosystems, affected by every visit from man. Please tread carefully, exercise caution (see pp. 13–14) and let your eyes, camera and magnifying glass be your main tools for exploring the seashore.

Understanding Tides

Tides are caused primarily by the gravitational forces of both the moon and the sun upon the earth. These gravitational forces override the centrifugal forces of the earth's rotation. They create a high tide, or "bulge" (see figures 1 and 2, p. 9) of water on the earth near the moon, which has a stronger gravitational effect than the sun because it is so much closer to the earth. A similar "bulge" is created on the opposite side of the earth. When the tide is high in one area, the displacement of water causes a low tide in another area. The earth makes one complete revolution under the "bulges" during one tide cycle, so there are two high tides and two low tides during each tide cycle. Tides have the greatest range when the moon is closest to earth.

During the days that immediately follow the new moon, the combined gravitational pull of the sun and the moon generate even higher tides and correspondingly lower tides (see Figure 1). During those days that follow the full moon, however, the moon's and sun's gravitational pull oppose each other (Figure 2), which dampens the tidal

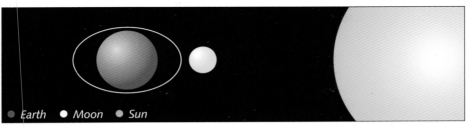
Figure 1

Figure 2

ffect. The lunar cycle is completed every 27$^{1/3}$ days, thus the moon orbits earth 13 times each year.

On each day of the year there are two high tides and two lows. The best time to view intertidal creatures is close to the lowest tide, so plan to arrive an hour or two before low tide. You can find this time—as well as the predicted height of the tide—by checking tide tables, available from tourist, sporting goods and marine supply stores and often published in local newspapers. (Keep in mind that these tables are usually based on standard time and on a particular geographical reference point, so daylight savings time and your actual location may have to be factored in.)

Tidal heights are measured from different reference points in the USA. For the most accurate information, use the reference point closest to the area you plan to visit. Tides of 0.0' are the average of the lower low tides for that year. Tides lower than this value are referred to as minus tides in the USA. Times when tide levels are lower than 0.0' are excellent for observing animal and plant life at intertidal sites. Any visit will be rewarding, but these are the optimum times to see intertidal life.

Low and high tide from the same location.

Understanding Intertidal Habitats

The rich marine life found "at the edge" of the Pacific Ocean is due in part to the wide variety of habitats in this range. Some creatures occupy quite a limited habitat, hardly venturing from a small area throughout most of their adult lives, because they can tolerate a very narrow range of conditions. Other more adaptable species can be seen in several intertidal zones and into the ocean depths.

The intertidal region comprises several different habitats and zones. Each combination provides a unique set of physical conditions in which many creatures survive and coexist.

An intertidal zone is characterized by several "key" species of marine flora and fauna—species typically found within that zone. The zone may be only a favored location; the species may occur in other zones as well.

Sand Beaches and Mud Flats

We often picture the Pacific coast as a vast sandy beach with gentle waves rolling toward shore, but this is only one of the many environments where seashore creatures have survived for centuries.

These two shorelines are very different, but they illustrate that intertidal species have preferred habitats (intertidal zones) regardless of the type of shoreline.

Thousands of years have passed since the last glaciers left their enormous deposits of sand and clay. Through time, the movement of land and sea have shifted huge volumes of these materials, which have provided numerous intertidal creatures with a place to burrow. The presence of many of these animals can be detected only by a slight dimple or irregularity in the surface of the sand or mud.

SAND BEACHES

Sand beaches are commonly found in both exposed and protected sites. Exposed sandy areas occur as sandspits or sand beaches. Creatures commonly seen on such beaches include the Pacific razor-clam (p. 140) and purple olive (p. 107). These and other species are well adapted to survive the surf-pounded beach. Protected beaches or sand flats, away from the pounding surf,

This sandy beach is home to the bay ghost shrimp (p. 171).

re a significantly different habitat, often occupied by Dungeness crab (p. 183), Nuttall's cockle (p. 137) and other species that are not adapted to the pounding waves of the outer coast. Some species occur in both exposed and protected sand beach habitats.

MUD FLATS

Mud flats are situated in sheltered locations such as bays and estuaries. Like sandy shores, they support a smaller variety of obvious intertidal life than rocky shorelines do. The fat innkeeper worm (p. 56) and Pacific gaper (p. 138) are species to look for in these areas.

Several species are characteristic of both mud flats and sand beaches. These include Lewis's moonsnail (p. 94), Pacific geoduck (p. 153) and softshell clam (p. 152).

This mud beach is home to several worms and clams.

Rocky Shores

Creatures have evolved special adaptations to live in certain habitats, so different species are found on exposed rocky shores than on sheltered ones. The California mussel (p. 125) and black Katy chiton (p. 70) occur in exposed areas, whereas more sheltered rocky sites harbor such creatures as the painted anemone (p. 43) and hairy hermit (p. 173).

Marine biologists divide rocky shores, as all shorelines, into several distinct intertidal zones: the splash zone and the high, middle and low intertidal zones. On rocky shores these zones are especially evident. The placement of creatures in the various zones is likely a complex combination of adaptations and environmental factors, including heat tolerance, food availability, shelter and suitable substrate availability. The presence of predators may also limit the range of intertidal zones an animal can inhabit. The purple star (p. 197), for example, preys upon the California mussel (p. 125), which pushes the mussel into a higher intertidal habitat.

Rocky shores occur in a wide variety of forms. These rocks move little from year to year.

This boulder area provides a habitat for a wide variety of life forms.

SPLASH ZONE
This zone can be easily overlooked as an intertidal zone, and the few small species present here seem to occur haphazardly. But these creatures are actually out of the water more than they are in it, so they must be quite hardy to tolerate salt, heat and extended dry periods. The acorn barnacle (p. 163) and ribbed limpet (p. 77) are two of these species.

HIGH INTERTIDAL ZONE
This zone is characterized by such species as the mask limpet (p. 76), aggregating anemone (p. 29) and California mussel (p. 125). Nail brush seaweed (p. 242) is one plant species that occurs in this zone, typically on top of rocks. (Seaweed species, like invertebrates, live in specific areas of the intertidal habitat.)

MIDDLE INTERTIDAL ZONE
This zone, also called the mid-intertidal zone, is home to the Vosnesensky's isopod (p. 167) and plate limpet (p. 76), as well as feather boa kelp (p. 231) and Pacific rockweed (p. 235). Most creatures in the mid-intertidal zone are normally not found in subtidal waters.

LOW INTERTIDAL ZONE
The sunflower star (p. 197) and purple sea urchin (p. 203) are among the many creatures to be found in the low intertidal zone, site of the most diverse and abundant marine life in the entire intertidal area. Creatures here often are found in subtidal waters too. In the low intertidal zone there is more food and shelter, and probably a greater chance that the animal will be caught in a very low tide, as low tides affect this zone only rarely during the year compared with the high and mid-intertidal zones. The time marine life is exposed to the heat of the sun is also reduced, so heat is not a major limiting factor on the creatures of the low intertidal zone. There are also more species to be found in subtidal waters.

Micro Habitats

UNDER ROCKS
This environment is an important one. Whether the shore is rock, gravel, sand or mud, many species such as the western spiny brittle star (p. 200), purple shore crab (p. 186) and black prickleback (p. 219) require this micro habitat for survival.

TIDEPOOLS
The grainyhand hermit (p. 172), mossy chiton (p. 68), umbrella crab (p. 174) and many other species are often found in tidepools but are not restricted to them. These creatures live in a somewhat sheltered environment that may be different from the zone in which the pool is located.

FLOATING DOCKS AND PILINGS
These man-made sites attract a wide range of marine plants and invertebrates. Like rocky shores, they provide solid places for settling. The plumose anemone (p. 34), Pacific blue mussel (p. 126) and shield-backed kelp crab (p. 179) commonly invade this habitat. Some are often attached to or living on the floating dock, so viewing is not restricted to low tides.

Harvesting Shellfish

One of the great pleasures of beachwalking can be gathering shellfish for a fresh dinner of seafood. Be aware that you need a license to harvest seashore life such as clams, oysters and (in some areas) seaweeds, and there are harvesting seasons and bag limits. Before you take any shellfish, check with local officials for current restrictions.

Shellfish harvest areas may also be closed due to pollution, or to harmful algal blooms such as red tides (see below). Check with local authorities to make sure the area you wish to harvest is safe. Then let the fun begin!

Red Tide

At certain times of the year, tiny algae reproduce rapidly in what is referred to as an algal bloom. Each of these algae can contain minute amounts of toxins, which are then concentrated in the body tissues of filter-feeding animals such as oysters, clams, mussels, scallops and other shellfish. Once the bloom dies, the animals' bodies begin to cleanse themselves of the toxins naturally, a process that takes time—as little as four to six weeks, but as long as two years for species such as butter clams.

Some experts believe that harmful algal blooms can produce a poison (saxitoxin) that is 10,000 times more toxic than cyanide. So if you eat even a tiny amount of shellfish that have ingested these toxins, you can become seriously or even fatally ill with paralytic shellfish poisoning (PSP). Symptoms include difficulty in breathing, numbness of tongue and lips, tingling in fingertips and extremities, diarrhea, nausea, vomiting, abdominal pain, cramps and chills. Reports of this ailment go as far back as human occupation along the Pacific coast.

Authorities regularly monitor shellfish for toxin levels, and affected areas are closed to shellfish harvesting. Watch for local postings of closures on public beaches and marinas, but to make sure, check with a PSP hotline or ask fisheries officials before harvesting any shellfish.

PSP (RED TIDE) HOTLINES

To obtain current marine toxin information, contact the following:
California Department of Health Services
"Shellfish Information Line" at 1-800-553-4133.
http://www.dhs.cahwnet.gov/

Protecting Our Marine Resource

Today more than ever it is essential for us to take responsibility for protecting our natural surroundings, including our marine environments. At many coastal sites human presence is becoming greater—sometimes too great. Habitat destruction, mostly from trampling, has been severe enough to cause authorities to close some intertidal areas to the public. In most cases this is not willful damage but people's lack of awareness of how harmful it can be simply to move around a seashore habitat.

To walk safely through an intertidal area, choose carefully where to step and where not to step. Sand and rock are always the best surfaces to walk on, when they are available. Mussels have strong shells that can often withstand the weight of a man without difficulty. Barnacles can also provide a secure, rough walking surface and can quickly recolonize an area if they become dislodged.

Please return all rocks carefully to their original positions, taking care not to leave the underside of any rock exposed. Take all containers back with you when you leave, as well as any debris from your visit. And please leave your dog at home when you visit intertidal sites.

Observing Intertidal Life

A magnifying glass is a must for any visit to the seashore, and a camera is the best way to take souvenirs. Another excellent item to take along is a clear plastic jar or plastic pail. Fill it with cool salt water and replace the water frequently. This will enable you to observe your finds for a short time with minimal injury to them. Make sure to return them to the exact spot where you found them. And if you must handle sea creatures, do so with damp hands so their protective slime coatings will not be harmed. Remember too that marine life in California is protected by law in marine reserves.

A Note of Caution

Before you visit an intertidal site, be aware of tide times and plan accordingly. During any visit to the beach it is important to stay out of low-lying areas that have no exit, and to keep a close watch on the water at all times. Many an unsuspecting beachcomber has become stranded on a temporary island formed by the incoming tide.

Strong wave action can take you by surprise. Dangerous waves come by a number of different names—sneaker waves, rogue waves, etc.—all of which indicate the nature of waves in exposed situations. Unexpected and powerful waves can and do take beach visitors from the shore. If you get caught off guard by a wave, the best defense is to lie flat, grabbing onto any available rocks that may provide a handhold. This will make it possible for the wave to roll over you rather than taking you out to sea. A vigorous surf can also toss logs up on shore unexpectedly. Please be careful!

Seaweeds can present a slippery obstacle to those venturing into intertidal areas. In order to provide food and protection for the many creatures found along the shore, these plants cover just about everything. In some areas a two-footed and two-handed approach is necessary to move around safely. Rubber boots with a good tread will help you observe intertidal life without slipping or getting soaked. It's a good idea to exercise caution around barnacles and such creatures, as their shells are hard and sharp-edged.

Even for a short visit, take along a backpack, some drinking water and a small first aid kit.

Visiting the intertidal sites of California is one of the most rewarding pastimes on earth. A little bit of preparation and a healthy dose of caution will help make every trip to the seashore a wonderful adventure.

Getting the Most Out of This Guide

The field guide section of this book (pp. 16–253) includes color photographs of the common animals and plants to be seen along the California seashore, and concise information that will help you identify species.

NAME: The current or most useful common name for the species; also the scientific name, a Latin name by which the species is known all over the world. This scientific name has two parts: the genus (a grouping of species with common characteristics) and the species.

OTHER NAMES: Any other common or scientific names known for the species.

DESCRIPTION: Distinguishing physical features, behavior and/or habitat to aid in identifying the species.

SIZE: Dimension(s) of the largest individuals commonly seen.

HABITAT: The type of area where the species lives (see Understanding Intertidal Habitats, pp. 10–12).

RANGE: The area of California where the species is found.

NOTES: Other information of interest, usually relating to the natural history of the species or ways in which it differs from a similar species.

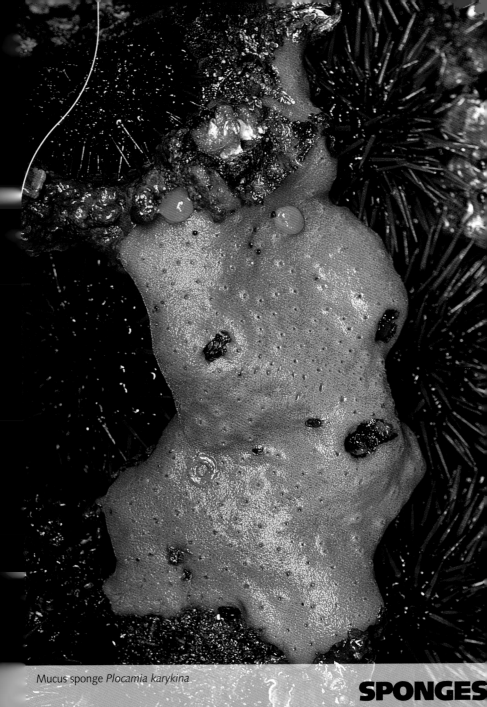

Mucus sponge *Plocamia karykina*

SPONGES
Phylum Porifera

Sponges are filter-feeding, colonial animals that live together as a larger unit. They appear to be plants, but are in fact invertebrate animals. Unique to the animal world, sponges have canals throughout their bodies that open to the surrounding water, allowing both oxygen and food particles to reach each sponge.

YELLOW BORING SPONGE
Cliona celata

OTHER NAMES: Sulfur sponge, boring sponge.
DESCRIPTION: A bright lemon-yellow body protrudes from a small $1/32-1/8$" (1-3 mm) hole in the shell of a variety of creatures.
SIZE: Lobes to $1/8$" (3 mm) in diameter.
HABITAT: From the low intertidal zone to water 400' (120 m) deep.
RANGE: Prince William Sound, Alaska, to Baja California, México.
NOTES: The boring sponge lives on the calcareous shells of a wide variety of sea life including large barnacles, some clams, moonsnails, oysters and others such as the giant rock scallop (see p. 131). This remarkable sponge bores holes in shells that are either living or dead. It secretes sulfuric acid to dissolve a small portion or pit in a calcareous shell. Under favorable conditions this sponge will overgrow its host completely.

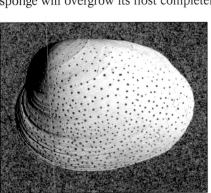

Holes bored by the yellow boring sponge.

SPONGES

BREAD CRUMB SPONGE

Halichondria panicea

OTHER NAME: Crumb of bread sponge.
DESCRIPTION: Soft, encrusting or crust-like sponge varying in color from yellow to light green. Several volcano-shaped oscula (pores) on surface.
SIZE: To 2" (5.1 cm) thick.
HABITAT: Low intertidal zone to subtidal depths of 200' (60 m).
RANGE: Bering Sea, Alaska, to Baja California, México.
NOTES: This species gets its name from bread crumb-like texture. If broken, it is said to smell like gunpowder after being ignited. Various nudibranchs feed on this sponge, including the Monterey dorid (see p. 118), which can often be found on the sponge once the tide has receded.

RED ENCRUSTING SPONGE

Ophlitaspongia pennata and others

OTHER NAMES: Red sponge, scarlet sponge, velvety red sponge.
DESCRIPTION: Bright red to red-orange with tiny, closely spaced star-like oscula (pores). Surface is velvety to touch.
SIZE: To 39" (1 m) in diameter, 1/4" (6 mm) thick.
HABITAT: On overhanging rocks and shady crevices, mid-intertidal zone to subtidal water 10' (3 m) deep.
RANGE: Sooke, Vancouver Island, BC, to near Puertocitos, Gulf of California.
NOTES: Most common of several species of sponges that look very similar to this species. A microscope is required to tell them apart. Check specimen closely to find the red nudibranch (see p. 117), which perfectly matches the sponge in color. The nudibranch also lays its red eggs on this sponge, its main food source.

Mucus Sponge
Plocamia karykina

OTHER NAMES: Smooth red sponge, red sponge.
DESCRIPTION: Scarlet to salmon-orange in color. This encrusting species is smooth and firm with oscula (pores) that are rather large, sparse and irregularly spaced. When the spicules are viewed under a microscope, they are observed as double-headed.
SIZE: To 10" (25 cm) wide and to 1" (2.5 cm) thick.
HABITAT: From the mid- to low intertidal zone.
RANGE: Vancouver, BC, to central Baja California, México.
NOTES: The mucus sponge is a common, thick species that produces large quantities of gelatinous mucus when disturbed. This species looks similar to several other intertidal species that may also be encountered. The others, however, are not known to produce large amounts of mucus.

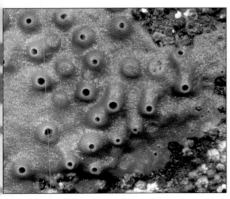

Purple Encrusting Sponge
Haliclona permollis

OTHER NAMES: Purple sponge, violet volcano sponge, encrusting sponge.
DESCRIPTION: **Soft** and encrusting or flat and crust-like sponge varying in color from pink to purple. Volcano-shaped oscula (pores) on surface.
SIZE: To 36" (91 cm) across, $1^{5/8}$" (4.1 cm) thick, but normally much smaller.
HABITAT: On rocks and floating docks and in tidepools, mid-intertidal zone to subtidal water 20' (6 m) deep.
RANGE: Northern southeast Alaska to southern California.
NOTES: Several nudibranchs feed on this beautiful sponge. The ringed nudibranch (see p. 120) is able to find this sponge by chemicals that the sponge releases into the water.

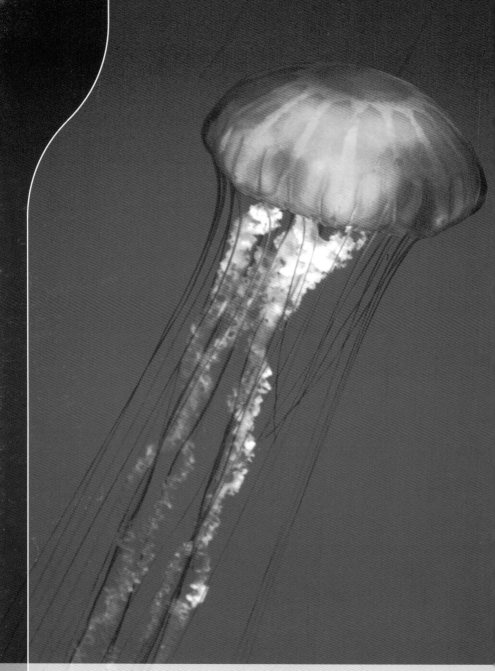

Sea nettle *Chrysaora fuscescens*.

HYDROIDS, JELLIES, SEA ANEMONES, COMB JELLIES
Phyla Cnidaria and Ctenophora

The phylum Cnidaria includes hydroids, scyphozoan (large bell-shaped) jellies, sea anemones and corals. This group has somewhat specialized organs for digesting or stinging. Many species have alternating generations from attached to free-swimming stages.

Comb jellies (phylum Ctenophora) are transparent like true jellies, but shaped much differently. These remarkable organisms are spherical or flattened into a ribbon shape, and they move by cilia or small hairs arranged in rows. Eggs and sperm are released from the mouth into the water, where they develop into small versions of the adult. There is no alternation of generations as there is in the bell-shaped jellies.

HYDROIDS, JELLIES, SEA ANEMONES
Phylum Cnidaria
Hydroids and Allies
Class Hydrozoa

Colony of clam hydroids attached to the shell of a bean clam.

CLAM HYDROID
Clytia bakeri

OTHER NAME: Medusae (sexual stage in its life history); formerly known as *Phialidium*.
DESCRIPTION: Colony is buff to gray in color and attached to living mollusks in tufts. Stalks have alternate branching on two sides.
SIZE: Stalks to 5" (13 cm) long but normally much smaller.
HABITAT: On exposed sandy beaches from the low intertidal zone to shallow subtidal waters.
RANGE: San Francisco, California, to Baja California, México.
NOTES: The clam hydroid is a common, easy-to-identify species. It is the only hydroid that attaches itself to living mollusks such as Pismo clams (see p. 151), bean clams (p. 144) and purple olives (p. 107). It attaches to the exposed end, which sits on the sand's surface, and it stays above the sand to feed and breathe. Every two or three years there is a population explosion and clam hydroids become extremely abundant.

HYDROIDS, JELLIES, SEA ANEMONES, COMB JELLIES

WINE-GLASS HYDROID
Obelia sp.

OTHER NAME: Sea plume.
DESCRIPTION: White in color. Individual main stems hold many smaller branches, giving an overall bushy look.
SIZE: To 10" (25 cm) tall.
HABITAT: Colonies are found attached to floats, large seaweeds or rock in shallow tidepools, low intertidal zone to water 165' (50 m) deep.
RANGE: Alaska to southern California.
NOTES: The wonderful name of this hydroid is very descriptive of the tiny coverings to the animal's feeding portions, but a microscope is necessary to see these wine-glass shapes and to make positive identification.

TURGID GARLAND HYDROID
Sertularella turgida

DESCRIPTION: Displays elongated yellow stalks with distinctive wavy pattern.
SIZE: 2" (5 cm) high, $1^{1/2}$" (4 cm) wide.
HABITAT: Low intertidal zone to water 528' (160 m) deep.
RANGE: BC to San Diego, California.
NOTES: Very little is known about the biology of this distinctive hydroid. We do know that to reproduce, it retains its eggs while sperm are released to the sea for fertilization.

Ostrich-Plume Hydroid
Aglaophenia sp.

Description: Elongated feather-like plumes attached to rock, varying in color from yellow to light red or black.
Size: To 5" (13 cm) high.
Habitat: In surf-swept rock clefts on exposed shores, intertidal to water 528' (160 m) deep.
Range: Alaska to San Diego, California.
Notes: Species is easily confused with a plant, but hydroids start out as larvae called planula, and as adults are composed of tiny filter-feeding polyps. The elongated yellow eggs are often found attached to the plumes.

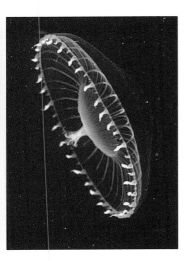

Water Jelly
Aequorea sp.

Other Names: Many-ribbed hydromedusa, many-ribbed jellyfish, water jellyfish, *Aequorea aequorea, Aequorea victoria*.
Description: Transparent bell-shaped jelly with 100 or more rib-like radial canals and trailing tentacles.
Size: To 3" (7.5 cm) in diameter.
Habitat: In open water and close to shore.
Range: Alaska to Baja California, México.
Notes: Various species of water jellies are found worldwide. Their luminescence is easily observed at night as soft, circular balls of pulsing light. This species is known to eat other species of jellies and on occasion to cannibalize its own species.

HYDROIDS, JELLIES, SEA ANEMONES, COMB JELLIES

BY-THE-WIND SAILOR
Velella velella

OTHER NAMES: Sail jellyfish, purple sailing jellyfish; formerly *Velella lata*.
DESCRIPTION: Bright blue float with a transparent triangular sail on the dorsal side.
SIZE: To $2^{1/2}$" (6 cm) long.
HABITAT: Normally on the ocean's surface but often found stranded on shore.
RANGE: Temperate and tropical oceans worldwide.

NOTES: By-the-wind sailor is a pelagic jelly that occasionally washes ashore by the hundreds in late spring and early summer. Its colorful base is made up of a float, which is comprised of gas-filled pockets. Its common name originates from the prominent sail that arises from the float. Although several tentacles surround this jelly's outer rim, they are harmless to man. This cosmopolitan species feeds on small fish eggs, copepods, etc.

Large Bell-shaped Jellies
Class Scyphozoa

The rhythmic pulses of jellies are intriguing to observe—indeed, their fluid movements have a near-hypnotic effect. The purpose of this movement is probably to keep the animal near the surface of the water. Its seemingly random wanderings are influenced and aided by water currents.

Jellies date back to Precambrian times: one Australian fossil has been aged at 750 million years. There are a thousand known species of these primitive carnivores, which feed primarily on zooplankton. The life cycle of a jelly has distinct stages, which include either a polyp (a tube-like organism with a mouth and tentacles to capture prey) or a medusa (umbrella-shaped organism) stage. The jelly captures its food, then lifts it to its mouth to eat.

Jellies are composed of as much as 96 percent water, but several species are consumed as food in various Pacific cultures. They are eaten boiled, dried or raw. The giant sunfish *Mola mola*, which has been known to grow to 2,700 lbs (1,215 kg), attains its huge size by feeding on jellies and similar coelenterates with nibbling and sucking techniques.

SEA NETTLE
Chrysaora fuscescens

DESCRIPTION: Color varies from reddish brown to yellowish brown. The oral arms in the centre are long and pointed. Bell has **24 tentacles** positioned on the edge in groups of 3.
SIZE: To 12" (30 cm) in diameter.
HABITAT: Near the shore.
RANGE: Gulf of Alaska to México but most common off Oregon and California.
NOTES: The sea nettle is a jelly with a very **unpleasant sting**. It is often found during fall and winter in aggregations near the shore or stranded on the beach. This jelly feeds on other jellies, comb jellies, fish eggs and larvae and a variety of other pelagic invertebrates. It swims continuously and is sometimes seen in fascinating displays at public aquariums.

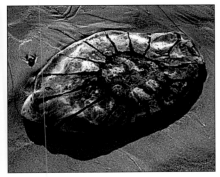

Stranded purple-striped jelly.

PURPLE-STRIPED JELLY
Pelagia colorata

OTHER NAMES: Purple banded jellyfish; formerly included with *Pelagia noctiluca* (*P. panopyra*).
DESCRIPTION: Overall color **silvery white with 16 purple bands** that radiate from a purple ring at the center of the bell. A series of oral arms and tentacles trail beneath.
SIZE: To 32" (80 cm) in diameter and several yards long.
HABITAT: Oceanic but occasionally stranded on shore or in tidepools.
RANGE: Bodega Bay, California, to San Diego County, California.
NOTES: The purple-striped jelly was once thought to be the same species as the much smaller and more widespread purple jelly *Pelagia noctiluca*. Scientists now recognize them as two distinct species. The purple-striped jelly is a large **toxic** species and should not be handled as it can deliver a painful sting. Its food consists of a wide variety of small marine life such as fish eggs and larvae, planktonic crustaceans and other jellies. The giant sunfish *Mola mola* is a predator of this large jelly.

HYDROIDS, JELLIES, SEA ANEMONES, COMB JELLIES

LION'S MANE
Cyanea capillata

OTHER NAMES: Sea blubber, sea nettle.
DESCRIPTION: Bell-shaped with smooth, near-flat top; shaggy clusters on ventral side with **150 fine trailing tentacles**. Yellow-brown to orange.
SIZE: Normally to 20" (50 cm) in diameter, tentacles to 10' (3 m) long; with one report of tentacles to 119' (36 m) long.
HABITAT: Usually found floating near surface, occasionally stranded on the beach.
RANGE: Alaska to southern California.
NOTES: The largest jelly in the world. Its tentacles deliver a burning sensation and rash to those who touch it. **Exercise caution**, even if you find a jelly stranded on the beach. All jellies are poisonous to some degree, and human reactions to the toxins vary from a mild rash to blistering and even, occasionally, death. Species feeds on small fish, crustaceans and other animals it comes into contact with. Some species of fish find the lower portion of the bell provides refuge from their enemies.

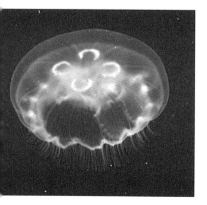

MOON JELLY
Aurelia labiata

OTHER NAMES: Moon jellyfish, white sea jelly.
DESCRIPTION: Whitish in color and translucent, often with a touch of pink, purple or yellow; bell-shaped with many short trailing tentacles and 4 round or horseshoe-shaped gonads or reproductive organs.
SIZE: 4–16" (10–40 cm) in diameter, 3" (7.5 cm) high.
HABITAT: Usually found floating near surface.
RANGE: Alaska to southern California.
NOTES: Sometimes collects in quiet waters such as harbors when plankton blooms in the spring. In these large aggregations, spawning also occurs. Tentacles cause a slight rash when handled. When the sun is visible, this jelly uses it as a compass to migrate in a southeasterly direction.
The moon jelly is often misidentified as *Aurelia aurita*, a European species now also found in California.

COMB JELLIES
Phylum Ctenophora

SEA GOOSEBERRY
Pleurobrachia bachei

OTHER NAMES: Comb jelly, cats eyes, sea walnut comb jelly.
DESCRIPTION: Transparent, egg-shaped organism with 2 long tentacles.
SIZE: To 5/8" (1.5 cm) in diameter; tentacles 6" (15 cm) long.
HABITAT: Found near shore, often in large numbers.
RANGE: Alaska to Baja California, México.
NOTES: This is the only species of comb jelly found off the waters of California. Comb jellies use sticky cells on the tentacles, rather than stinging cells, to capture food. They are often found in spring and summer swimming in large swarms. Occasionally they wash up on the beach. Each individual is both male and female. Eggs and sperm are released from the mouth to be fertilized in open water. This species has been called a voracious carnivore, as swarms can severely reduce schools of young fish. Most comb jellies are bioluminescent, but this species cannot produce its own light.

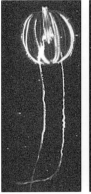

Left: with extended tentacles.
Right: with retracted tentacles.

Phylum Cnidaria

Sea Anemones and Cup Corals
Class Anthozoa

Sea anemones, like their relatives the jellies, possess nematocysts (stinging cells), primarily on their tentacles. The animal uses its nematocysts to sting its prey when contacted. If a person's hand touches the tentacles they only feel sticky, since the skin on our hands is too thick to allow penetration, but a stinging sensation was definitely felt by an individual who licked the tentacles of one species, using his much more sensitive tongue. (This technique is, however, not recommended!) Most anemones feed upon fishes, crabs, sea urchins, shrimps and similar prey. The habitat varies from the high intertidal zone to as deep as 30,000' (9,000 m) for some species in the Philippines.

The class Anthozoa includes both sea anemones and cup corals, anemone-like organisms with hard, cup-shaped skeletons.

HYDROIDS, JELLIES, SEA ANEMONES, COMB JELLIES

ORANGE CUP CORAL
Balanophyllia elegans

OTHER NAMES: Orange-red coral, solitary coral.
DESCRIPTION: Bright orange, cup-shaped organism with hard outer seat-like shape surrounded by many small tentacles. Resembles a small anemone; its hard seat helps in identifying it correctly as a hard coral.
SIZE: To 1/2" (1.2 cm) in diameter.
HABITAT: On shady sides and under rocks or boulders, low intertidal zone to open water 70' (21 m) deep.
RANGE: BC to Baja California, México.
NOTES: Identification of this species is made easy by the fact that no other orange stony coral is found intertidally in California. Its vivid orange color comes from a fluorescent pigment. Feeding is accomplished with small, transparent-looking tentacles and a mouth that opens to trap food. Although this species is small, its presence is always a welcome splash of color.

MOONGLOW ANEMONE
Anthopleura artemisia

OTHER NAMES: Green burrowing anemone, buried sea anemone, beach sand anemone.
DESCRIPTION: Overall coloration varies from brown to gray or olive-green. Slender, tapering, bright pink, orange, green or blue **tentacles with distinctive white bands**. Column is usually covered with sand and bits of shell.
SIZE: To 2" (5 cm) in diameter, with only disc and tentacles protruding above the surface.
HABITAT: In sand or gravel on sheltered cobblestone or rocky shorelines, mid- or low intertidal zone and into subtidal, in both exposed and sheltered locations.
RANGE: Alaska to southern California.
NOTES: The species gets its name from the luminous quality often exhibited by the tentacles. The moonglow anemone is usually found in a sandy area, attached to a large shell or rock, buried up to 12" (30 cm) beneath the surface. The aggregating anemone (see opposite) can look similar when living in sand, but it lacks the white bands on its tentacles.

AGGREGATING ANEMONE
Anthopleura elegantissima

OTHER NAMES: Clustering aggregate anemone, pink-tipped green anemone, surf anemone.

DESCRIPTION: Coloration variable. The anemone displays **pale green tentacles with pinkish or purple tips** and a pale green to gray column.

SIZE: Isolated individuals may grow to 10" (25 cm) in diameter, 20" (51 cm) high. Aggregating individuals, however, reach less than a third of this size.

HABITAT: In colonies attached to rocks above the low tide line and in tidepool situations on exposed and protected shores with active currents. The presence of sand seems be important in where this species colonizes.

RANGE: Alaska to Baja California, México.

NOTES: Aggregating anemones are well known for their ability to multiply asexually by dividing into two identical individuals. As a result, they are capable of colonizing large rock surfaces as genetic clones. At some locations, their clones carpet rock surfaces entirely. Sexual reproduction is also possible.

By necessity, this species must be very tolerant of harsh conditions including exposure to the sun, wind and waves. Individuals are often observed in a closed position, out of water, which prevents them from drying out. However, this changes their appearance dramatically.

This anemone's predators include the shag-rug nudibranch (see p. 123) and the leather star (p. 194).

Closing disc in a tidepool.

HYDROIDS, JELLIES, SEA ANEMONES, COMB JELLIES

GIANT GREEN ANEMONE
Anthopleura xanthogrammica

OTHER NAMES: Green anemone, rough anemone, solitary anemone.

DESCRIPTION: Disc and tentacles are a beautiful emerald green, short column is olive-brown. Sand and shell fragments are commonly found in disc.

SIZE: To 12" (30 cm) in diameter, 12" (30 cm) high.

HABITAT: On exposed rocky shores, intertidal to water deeper than 50' (15 m).

RANGE: Alaska to Panama.

NOTES: This sea anemone is found in both a solitary existence and in groups, often in tidepools. Microscopic green algae live inside the tentacles, giving the animal its green color. This species attains adult size at age 14 or 15 months and has been known to live longer than 30 years in captivity.

The giant green anemone has been utilized by humans in different ways. The Aboriginal peoples cooked it carefully over a fire before eating it. More recently, this species has been used as the source for a heart stimulant to vertebrates.

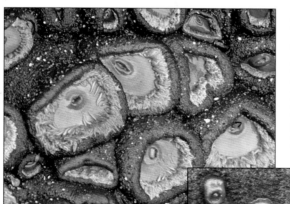

A group, out of water with closed discs.

Proliferating Anemone
Epiactis prolifera

OTHER NAMES: Brooding anemone, small green anemone.
DESCRIPTION: Green, brown or red with **white pinstripes on base, that do not extend to center of column**. Young of various sizes are usually present, up to 30 at one time crowded into a single row.
SIZE: Normally to 1 1/4" (3 cm) in diameter, 1" (2.5 cm) high.
HABITAT: In unprotected sites with eel-grass, kelp beds and the sides of rocks providing habitat, high tide line to 30' (9 m).
RANGE: Alaska to La Jolla, San Diego County, California.
NOTES: This anemone has a sex life similar to that of orchids, referred to as gynodioecy. Young adults are usually female. As mothers, they grow in size and they also grow testes. Upon full maturity they are hermaphrodites, having both testes and ovaries.

The eggs develop within the digestive cavity of the parent, exit through the mouth and eventually attach themselves to the middle of the parent's column. Later the young move away from the parent to venture forth on their own.

Striped Anemone
Haliplanella luciae

OTHER NAME: Formerly *Haliplanella lineata*.
DESCRIPTION: Olive green overall, with pale orange or yellow stripes running up and down the column.
SIZE: To 3/4" (1.9 cm) high, 1/4" (6 mm) in diameter.
HABITAT: On rocks and similar objects in shallow water, or salt marshes in high intertidal zone.
RANGE: Southern BC to southern California.
NOTES: This anemone is thought to have been introduced accidentally in the late nineteenth century, when the Pacific oyster (see p. 130) was brought to North America for commercial purposes. The striped anemone's small size makes it an easy species to overlook. Like other species, this one has specialized tentacles called catch tentacles, which can be used in disputes with other anemones of the same or different species. These battles can be fatal to one or the other anemone.

HYDROIDS, JELLIES, SEA ANEMONES, COMB JELLIES

RED-BEADED ANEMONE
Urticina coriacea

OTHER NAMES: Beaded anemone, leathery anemone, stubby rose anemone; formerly *Tealia coriacea*.

DESCRIPTION: Short red and gray tentacles with white bands. Thick column varies in color from brownish red to hot pink with thick sucker-like tubercles covered by sand and shell debris. As a result, column is normally hidden from view.

SIZE: To 4" (10 cm) in diameter, 5 1/2" (14 cm) high.

HABITAT: Usually found half-buried in sand or gravel or situated in gravel-filled crevices, low intertidal zone to water 50' (15 m) deep.

RANGE: Alaska to south of Carmel, Monterey County, California.

NOTES: This anemone attaches itself to rocks or other solid objects beneath the surface, leaving only its crown visible. It is commonly found in shallow, rocky intertidal areas. Its chief enemy is the leather star (see p. 194). Little else is known about this anemone's biology.

The disc of this red-beaded anemone is open,

while this disc is closing in a tidepool.

This individual has temporarily turned its stomach inside out.

Painted Anemone
Urticina crassicornis

OTHER NAMES: Red and green anemone, Christmas anemone, northern red anemone, dahlia anemone, mottled anemone; formerly *Tealia crassicornis*.

DESCRIPTION: Column varies considerably from green with red blotches to solid color range of light yellow-brown, white, green or occasionally totally red. Small wart-like cells may or may not be present on column. Color of tentacles generally similar to or somewhat lighter than column.

No debris attached to column of this species.

SIZE: Normally to 4" (10 cm) but can reach 10" (25 cm) in diameter; to 10" (25 cm) high.

HABITAT: In protected sites such as under ledges or vertical rock surfaces, between low intertidal and subtidal zones.

RANGE: Alaska to south of Carmel, Monterey County, California.

NOTES: The painted anemone has been known to live to 80 years in captivity. Young individuals are normally found higher intertidally, while occasionally the tidepooler can view older, larger specimens at the lowest of tides. These large individuals have also been described by some observers as "obscene" for the manner in which they may hang from their substrate. This circumpolar species is referred to as the thick-petaled rose anemone along the east coast.

A typically colored individual.

HYDROIDS, JELLIES, SEA ANEMONES, COMB JELLIES

PLUMOSE ANEMONE
Metridium senile

OTHER NAMES: Sun anemone, frilled anemone, powder puff anemone, white-plumed anemone.
DESCRIPTION: Smooth white, yellow, orange or brown in color, with fewer than 100 tentacles.
SIZE: Normally to 2" (5 cm) high but can reach 4" (10 cm); **to 2" (5 cm) in diameter at base**.
HABITAT: Common in protected waters attached to hard objects such as wharves, dock pilings and rocks.
RANGE: Southern Alaska to southern California.

White form on a dock, underwater.

NOTES: This anemone is easily found as it is commonly attached to floating docks and similar sites throughout its range. It feeds on copepods and various invertebrate larvae. This anemone reproduces asexually; in fact, a new anemone can arise from tissue left behind when this anemone moves along to a new site. The new individual is a clone of the original.

Closed (left) and open (right) individuals underwater.

Above the water.

STRAWBERRY ANEMONE

Corynactis californica

OTHER NAMES: Pink anemone, club-tipped anemone.

DESCRIPTION: Color of tentacles varies considerably from orange to pink, scarlet or red; occasionally yellow, brown or even purple tentacles may be found. A **white club graces the tip of each tentacle**.

SIZE: To 1" (2.5 cm) in diameter.

HABITAT: On rocks and in tidepools from the low intertidal zone to water 99' (30 m) deep.

RANGE: BC to San Martin Island, Baja California, México.

NOTES: This beautiful species forms clones similar in nature to the aggregating anemone (see p. 29). As a result, identical looking anemones may cover large areas of a rock. The strawberry anemone is not a true anemone but rather closely related to the stony corals; however, it lacks the calcareous exoskeleton of the coral. The club-like tips on their tentacles, which are the largest stinging cells known to scientists, are used to capture a wide variety of small marine animals.

Tubes of the sand-castle worm
Phragmatopoma californica

MARINE WORMS
Phyla Platyhelminthes, Nemertea, Annelida, Sipuncula and Echiura

Marine worms are a collection of unrelated yet similar animal groups. These worms are classified in several phyla, including flatworms, ribbon worms, segmented worms, peanut worms and echiuran worms.

FLATWORMS
Phylum Platyhelminthes

These unsegmented worms are characteristically flat and do not have blood or circulatory systems. Flatworms may have eyespots, which are not restricted to the head region. These organs merely detect the presence of light.

LARGE LEAF WORM
Kaburakia excelsa

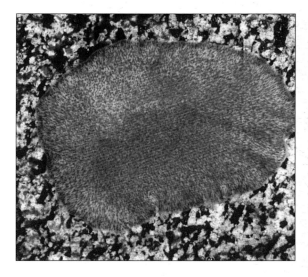

OTHER NAME: Leaf worm.
DESCRIPTION: Color varies from orange to brown. Tough, firm, oval body. 2 short tentacles, each with an eyespot, near brain.
SIZE: To 4x2$^{3/4}$" (10x7 cm), $^{1}/_{8}$" (3 mm) thick.
HABITAT: Sometimes found around mussel beds or seaweeds and under rocks, sometimes in large numbers; intertidal zone.
RANGE: Sitka, Alaska, to Newport Harbor, California.
NOTES: This giant species responds negatively to sunlight. Approximately 50 minute light-sensitive eyespots are located along the margin. This worm glides over the substrate by using thousands of tiny cilia, blending in to its environment so well that it is often difficult to detect. There is a mouth on the lower or ventral side but no anus, thus wastes pass back through the worm's mouth to be expelled. It is believed that this species feeds on animals such as limpets.

MARINE WORMS

TAPERED FLATWORM
Notocomplana acticola

OTHER NAMES: Common flatworm, brown flatworm; formerly *Notoplana acticola, Leptoplana acticola*.
DESCRIPTION: Tan overall with darker center patches on the dorsal side. The body is widest at the front, and **tapers to a point at the rear**. Two round clusters of eyespots are present with no obvious tentacles.
SIZE: To 2 3/8" (6 cm) long by 3/4" (1.9 cm) wide but normally much smaller.
HABITAT: Under rocks from the high to low intertidal zone.
RANGE: Entire California coast.
NOTES: The tapered flatworm appears to move like a floating carpet by using its tiny cilia. It is a predator, feeding on various mollusks and crustaceans up to half its size. Foods include the ribbed limpet (see p. 77) and in captivity it has been observed to feed on the red nudibranch (p. 117). This common flatworm is a hermaphrodite: in most cases both gonads, containing mature sperm, and eggs are present.

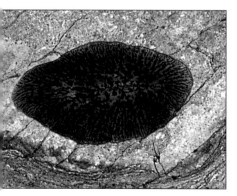

OVAL FLATWORM
Pseudoalloioplana californica

OTHER NAMES: Formerly *Planocera californica, Alloioplana californica*.
DESCRIPTION: Oval body, greenish in color with brown zigzag patches radiating to the outer edge. Eyespots cover the nipple-like contractable tentacles. Mouth is central on the underside.
SIZE: To 1 1/2" (4 cm) long and 3/4" (1.9 cm) wide.
HABITAT: In crevices or under rocks and boulders resting on sand or gravel in the mid-intertidal zone.
RANGE: California to Baja California, México.
NOTES: Tiny snails are one of the known foods of the oval flatworm. The zigzag markings on this flatworm indicate the extensive branching of its digestive tract.

LONG SPECKLED FLATWORM
Enchiridium punctatum

DESCRIPTION: White to cream or greenish in color, with small brown to black spots scattered over an elongated body.
SIZE: To 1 1/2" (4 cm) long.
HABITAT: On rocky shores from the low intertidal zone to water 50' (15 m) deep.
RANGE: Redondo, Los Angeles County, to Gulf of California, México.
NOTES: The long speckled flatworm has two cerebral eye clusters, each comprising about 20 tiny eyes, as well as a series of tiny eyes circling the edge of its body. Its sucker is found on the posterior end of its underside. This beautiful flatworm is found hidden between rocks in crevices and similar situations.

RIBBON WORMS
Phylum Nemertea

The worms in this group are more advanced than flatworms, with blood or circulatory systems. The ribbon worm has a retractable proboscis (snout) with either sticky glands or poisonous hooks to capture its prey.

SIX-LINED RIBBON WORM
Carinella sexlineata

OTHER NAMES: Six-lined nemertean, lined ribbon worm; formerly *C. dinema*, *Tubulanus sexlineatus*.
DESCRIPTION: Chocolate brown or black in color, with **5–6 white longitudinal stripes** and up to 150 cross-bands.
SIZE: Normally to 8" (20 cm) but occasionally to more than 39" (1 m) long.
HABITAT: Among rocks, mussels, algae and pilings, from the low intertidal zone to water 26' (8 m) deep.
RANGE: Sitka, Alaska, to southern California.
NOTES: The six-lined ribbon worm is a beautiful species that builds a transparent, parchment-like tube in which it then lives. A variety of worms make up the diet of this distinctive nemertean.

MARINE WORMS

ORANGE RIBBON WORM
Tubulanus polymorphus

OTHER NAMES: Orange nemertean, red ribbon worm; formerly *Carinella rubra*.

DESCRIPTION: Vivid red or orange. Elongated round, soft body with no distinct markings on the head.

SIZE: To 3' (90 cm) long, and to an amazing 10' (3 m) in large specimens, but often somewhat contracted due to its elastic nature. To $3/16$" (5 mm) wide.

HABITAT: Among mussels, in gravel or under rocks in quiet areas, sometimes in higher numbers, rocky low intertidal zone to 165' (50 m) deep.

RANGE: Aleutian Islands, Alaska, to San Luis Obispo County, California.

NOTES: This highly visible worm can often be observed moving slowly along in intertidal pools, hunting for prey. Its slow movement and rounded shape are characteristic. Often common in spring and summer.

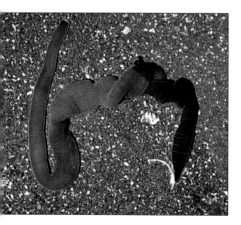

ROSE RIBBON WORM
Cerebratulus montgomeryi

DESCRIPTION: Overall color is dark rosy pink, with **a white, mouth-like mark across the tip of the head**.

SIZE: To 80" (2 m) long.

HABITAT: Under rocks in mud, from the low intertidal zone to water 1,320' (400 m) deep.

RANGE: Unalaska Island, Alaska, to California.

NOTES: The rose ribbon worm is a large species that does not break easily when handled, unlike other ribbon worms, which break or constrict into several pieces when they are picked up. This is believed to be a way in which they multiply: each worm piece regenerates its missing parts to form a whole new identical worm.

Velvet Ribbon Worm
Lineus pictifrons

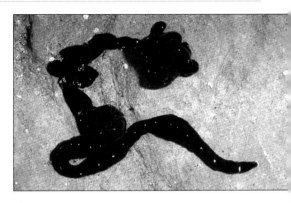

Description: Color varies widely from deep brown to chestnut, slate grey or bluish; usually with a **velvety sheen**. A pair of orange spots are usually found at the anterior end of the head. **Yellow, diamond-shaped points** are normally positioned along a median line. Yellow rings and longitudinal markings are also often positioned along the body, but this varies with individuals. Body generally twisted and snarled overall.
Size: Normally to 6" (15 cm) long but occasionally to 20" (50 cm).
Habitat: From the intertidal zone to water 130' (40 m) deep.
Range: California to Gulf of México.
Notes: The velvet ribbon worm is found in a wide range of habitats, including under rocks and in rock crevices, kelp holdfasts as well as among algae, tunicates or worm tubes. It may also be found in the muddy bottoms of bays and harbors.

Two-Spotted Ribbon Worm
Amphiporus bimaculatus

Other Names: Chevron amphiporus, thick amphiporus.
Description: Orange or brownish red; short, stocky body. Lighter-colored **head has a pair of dark triangular markings**.
Size: To 5" (13 cm) long.
Habitat: Usually in a rocky situation, low intertidal zone to water 450' (137 m) deep.
Range: Sitka, Alaska, to Baja California, México.
Notes: This common species, which resembles a leech, can sometimes be seen swimming after it has been disturbed. It breeds in July in the southern portion of its range. Like many worms, this one shuns light. The 2 eye-like markings in the head region are not eyes; tiny light-detecting organs located near these markings help the worm detect light and find its way to darkness.

MARINE WORMS

PINK-FRONTED RIBBON WORM
Amphiporus imparispinosus

OTHER NAME: Thin amphiporus.
DESCRIPTION: Color is white overall, often with a **pinkish or orange tinge toward its anterior end**. The body of this small worm is flattened, elongated and elastic in appearance.
SIZE: Normally to 6" (15 cm) long.
HABITAT: On the open coast in rocky areas of the intertidal zone.
RANGE: Bering Sea to Ensenada, Baja California, México.
NOTES: It is believed that the pink-fronted ribbon worm feeds on amphipods while it moves between mussels or barnacles, or around seaweed holdfasts. The white intertidal ribbon worm *Amphiporus formidabilis* is a similar species that lacks the pink or orange coloration and reaches a larger size, but a microscope is needed to observe other differences in the proboscis (snout), which is its retractable feeding apparatus.

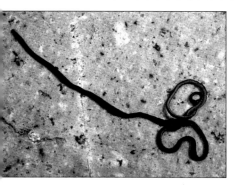

GREEN RIBBON WORM
Emplectonema gracile

OTHER NAMES: Green and yellow ribbon worm; formerly *E. viride, Nemertes gracilis*.
DESCRIPTION: Dark green above and yellowish green below. The body is flattened, long and elastic in appearance.
SIZE: To 4" (10 cm) long and $1/16$" (2 mm) wide.
HABITAT: Rocky shores and among barnacles and mussels in the mid-intertidal zone.
RANGE: Aleutian Islands, Alaska, to Baja California, México.
NOTES: The green ribbon worm is a small, distinctive species with many eyespots on the sides of its head. A group of 8 to 10 eyespots can be seen on the front of each side; these are best viewed from below. Another cluster of 10–20 smaller eyespots can be found above the brain, but these are more difficult to see. This nemertean is most often active at night feeding on barnacles, including the acorn barnacle (see p. 163). It is a specialist that uses its proboscis (snout) to suck out the flesh of its prey. There are also reports of this worm feeding on segmented worms and limpets.

PURPLE-BACKED RIBBON WORM
Paranemertes peregrina

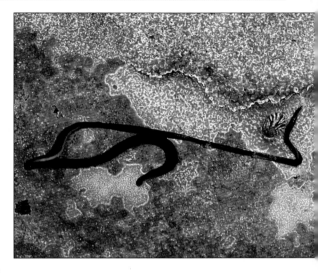

OTHER NAMES: Purple ribbon worm, restless ribbon worm, mud nemertean, wandering ribbon worm, wandering nemertean.
DESCRIPTION: Purple-brown varying to bluish purple above, and cream-colored below. This is a long, flattened worm, elastic in appearance.
SIZE: Normally to 5" (13 cm) long and 3/16" (5 mm) wide. Occasionally to 10" (25 cm) long.
HABITAT: In rocky and muddy areas from the mid- to low intertidal zone.
RANGE: Aleutian Islands, Alaska, to Ensenada, Baja California, México.
NOTES: The purple-backed ribbon worm feeds on a variety of segmented worms, including the pile worm (see p. 45). This species cannot detect its prey until it actually bumps into it, at which time it immediately coils its proboscis (snout), which is an extendable feeding apparatus, around its unlucky prey. The prey is then stabbed with a stylet, a nail-like weapon on the tip of the worm's proboscis, and a nerve poison called anabaseine is released to paralyze the prey. The worm then briefly lets go, allowing the neurotoxin to work before eating the prey whole. The purple-backed ribbon worm is known to live as long as $1^{1/2}$ years.

SEGMENTED WORMS
Phylum Annelida

These worms are easily identified by the many visible rings that make up their bodies. Over 9,000 species of segmented worms have been identified. Many marine species are found on the sea bottom but are not restricted to it.

MARINE WORMS

PROBOSCIS WORM
Glycera sp.

OTHER NAMES: Beak thrower, bloodworm.
DESCRIPTION: Light-colored, iridescent body, long and earthworm-like. Elongated head resembles a tapered point, which can fire out a proboscis (snout) containing 4 hooks at the tip.
SIZE: To 14" (35 cm) long.
HABITAT: In mud and muddy sand in areas of eel-grass and under rocks, low intertidal zone to water 1,040' (315 m) deep.
RANGE: BC to Baja California, México.
NOTES: 4 black jaws grace this worm's proboscis, which it can evert rapidly to almost a third of its body length. The worm uses its proboscis to capture prey. It also extends the proboscis into the sand, where the end swells to act as an anchor in bringing the body forward. Please **exercise caution** if you handle this worm. Its remarkable proboscis can inflict quite a bite.

BAT STAR WORM
Ophiodromus pugettensis

OTHER NAME: Formerly *Podarke pugettensis*.
DESCRIPTION: Color varies from reddish brown to purple or black, with several visible appendages.
SIZE: To 1 1/2" (3.8 cm) long.
HABITAT: From the low intertidal zone to subtidal waters.
RANGE: Pacific coast.
NOTES: This worm lives among the tube feet of the bat star (see p. 193) and other sea stars as a commensal worm (it doesn't harm its host). The bat star worm locates its host by scent, and studies show that nearly half of them leave their host each day and return later to another. There have been instances where up to 20 worms have been found living on one bat star at a time. This number is much higher during winter months. The bat star worm can also be found free-living on silty bottoms, floats and pilings.

Pile Worm
Nereis vexillosa

Other Names: Mussel worm, clamworm.
Description: Males are iridescent blue-green, females are dull green. Body is made up of approximately 200 segments, each of which has a pair of leg-like appendages. Species makes a strong first impression, thanks to pincer-like claws gracing the proboscis (snout).
Size: To 12" (30 cm) long.
Habitat: Found in a wide variety of habitats, including sand and mud beaches, beneath rocks, on wharf pilings and in mussel beds, in **high to mid-intertidal zones** of protected shores to exposed shorelines. The pile worm builds a loose, flexible tube from a mucus secretion binding sand and stones together.
Range: Alaska to San Diego, California.
Notes: The breeding season of this common species is linked to the full moons of summer, at which time huge congregations can be observed at night using their flattened leg-like appendages for swimming. The males release their sperm, then the females release eggs. Once the breeding sequence is completed, both sexes perish. Please **exercise caution** when handling the pile worm. It has been known to deliver the occasional nasty bite!

Shimmy Worm
Nephtys sp.

Other Name: Sandworm; sometimes misspelled *Nephthys*.
Description: Color varies. Body is beautifully iridescent. **No long tentacles in head region. Leg-like parapodia (paddles) look fuzzy** because of numerous hairs.
Size: To 12" (30 cm) long.
Habitat: Burrows in sand or sand-mud, mid-intertidal zone to subtidal waters 2,875' (869 m) deep.
Range: Alaska to Peru.
Notes: This rapid burrower everts a unique muscular proboscis (snout) to feed on worms and crustaceans. Like several other worms, it swims with a side-to-side wriggling movement aided by the paddling motions of its parapodia. A microscope is necessary to identify this species positively.

MARINE WORMS

CALIFORNIA FIREWORM
Pareurythoe californica

DESCRIPTION: Color is pinkish overall with red and purple highlights. The leg-like parapodia (paddles) are covered in white, glistening hair-like setae.
SIZE: To 4 1/2" (11 cm) long.
HABITAT: In sand or mud from the low intertidal zone to shallow subtidal waters.
RANGE: Central to southern California.
NOTES: The California fireworm is a cousin of the well-known tropical fireworms (family Amphinomidae). This beautiful species should not be touched: it protects itself with its hair-like setae, which contain an irritating poison. These setae are easily broken off as they penetrate the skin, and the poison leaves a burning sensation that can last a day or two.

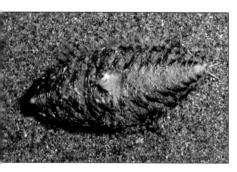

COPPER-HAIRED SEA MOUSE
Aphrodita japonica

OTHER NAMES: Sea mouse; formerly *Aphrodite japonica*; possibly includes *A. refulgida*.
DESCRIPTION: Color of body is light brown overall. The body is oval and comprises about 40 segments covered in hair-like setae. The **setae** have an impressive **copper or gold-colored** metallic sheen.
SIZE: To 9" (22 cm) long, but normally much smaller.
HABITAT: In soft sediments of subtidal waters.
RANGE: Alaska to Ecuador.
NOTES: This plump, unusual looking worm is sometimes found washed up on the sand after a storm. Its scientific name *Aphrodita* is appropriate: Aphrodite was a Greek goddess of love and beauty who was tossed onto the shore after a storm. The copper-haired sea mouse is also found in Japan, as its scientific name indicates. The hair-like setae are very sharp and have been known to deliver painful wounds. 2 pairs of minute eyes are present but difficult to see.

Fragile Scaleworm
Arctonoe fragilis

OTHER NAME: Frilled commensal scaleworm.
DESCRIPTION: Light in color, often matching color of host. 2 rows of ruffled and folded edges.
SIZE: To 3 1/2" (8.5 cm) long.
HABITAT: Low intertidal zone to water 908' (275 m) deep.
RANGE: Alaska to San Francisco, California.

NOTES: This worm feeds on detritus. It is often found on the underside of several species of sea stars. Normally only one fragile scaleworm is found on a host at a time, but there have been reports of up to 4 at a time being found on a host.

Red Commensal Scaleworm
Arctonoe pulchra

OTHER NAME: Scale worm.
DESCRIPTION: Brick red. Each scale may or may not display a single dark spot.
SIZE: To 2 3/4" (7 cm) long.
HABITAT: In the cavity of several intertidal hosts (see Notes).
RANGE: Gulf of Alaska to Baja California, México.

NOTES: This species, like several other scaleworms, feeds on detritus. It is found on the underside of a variety of organisms, including the rough keyhole limpet (see p. 73), giant Pacific chiton (p. 69), leather star (p. 194) and California sea cucumber (p. 205).

MARINE WORMS

RED-BANDED COMMENSAL SCALEWORM
Arctonoe vittata

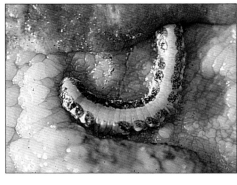

OTHER NAMES: Scale worm; formerly *Polynoe vittata*.
DESCRIPTION: Pale yellow with a red-brown band; series of scales along the edges of the back.
SIZE: To 4" (10 cm) long.
HABITAT: Low intertidal zone to water 800' (244 m) deep.
RANGE: Pacific coast.

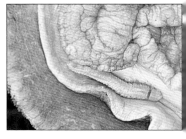

NOTES: Some individuals are free living, some live commensally. Hosts vary from the rough keyhole limpet (see p. 73) and giant Pacific chiton (p. 69) to 9 species of sea stars, including the leather star (p. 194) and Pacific blood star (p. 195). The host attracts the worm by releasing a chemical scent. This species actually helps protect its host from predators such as the purple star (p. 197) by biting at the predator's tube feet when it attacks. It is definitely in this worm's best interest to ensure its host continues to live!

BLOODWORM
Euzonus sp.

OTHER NAME: Formerly *Thoracophelia*.
DESCRIPTION: Bright red, because of hemoglobin in the blood.
SIZE: To 1 1/2" (4 cm) long.
HABITAT: On sand beaches around the mid-intertidal zone, burrowing to a depth of 4–12" (10–30 cm) below the surface.
RANGE: Vancouver Island, BC, to Baja California, México.

NOTES: Bloodworms are often found in very high numbers, frequently in a band less than 3' (1 m) wide along the shore. They burrow deeper into the sand as the water slowly recedes and the sand dries out. These worms are very important food for a wide variety of migrating shorebirds.

Pacific Lugworm
Abarenicola pacifica

OTHER NAME: Lugworm.
DESCRIPTION: Light orange with branched gills along the body (color is best viewed under water).
SIZE: To 6" (15 cm) long.
HABITAT: On sand-mud beaches, high intertidal zone.
RANGE: Alaska to Humboldt Bay, California.
NOTES: Telltale castings of mud and sand often indicate the presence of this species. Beneath the slender fecal casting lies a J-shaped burrow harboring the lugworm,

The distinctive sand casting

which extracts bacteria and organic debris from the sand. When this worm is first uncovered, its long body is greenish in color, but once immersed in water, this seemingly unattractive creature contracts and quickly changes to a pleasant-looking worm with the hemoglobin of its body becoming visible. If left long enough, the gills eventually expand and become bright red. Lugworms have been used as bait by fisherman in Europe and elsewhere.

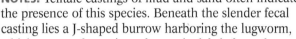

Red-banded Bamboo Worm
Axiothella rubrocincta

OTHER NAMES: Red-banded tube worm, bamboo worm, jointworm, *Clymenella rubrocincta*.
DESCRIPTION: Greenish body with red bands on segments. This worm produces a long, narrow, sandy tube that projects about 1" (2.5 cm) above the sand when the tide is out.
SIZE: To 8" (20 cm) long and 1/4" (6 mm) wide.
HABITAT: In muddy sand in the intertidal zone. Commonly found in bays and estuaries.
RANGE: Southern Alaska to México.
NOTES: Bamboo worms get their name from their elongated segments, which bear a striking resemblance to bamboo cane. The red-banded bamboo worm lives in a U-shaped tube that extends into the sand to 12" (30 cm) deep. Sand is swallowed in order to feed on detritus and other organic materials.

MARINE WORMS

SAND-CASTLE WORM
Phragmatopoma californica

OTHER NAMES: Pacific black-bristled honeycomb worm, black-topped honeycomb worm, colonial sand tube worm, seabee worm.

DESCRIPTION: A cream-colored worm with lavender tentacles and **black bristles**. This worm is a **tube** builder and is almost always found in a colony that **forms a honeycomb design** (see p. 36).

SIZE: To 2" (5 cm) long.

HABITAT: On rocky shores with sand nearby from the low intertidal zone to water 245' (75 m) deep.

RANGE: Central California to Ensenada, Baja California, México.

NOTES: Tiny sand grains are cemented together by sand-castle worms into a distinctive honeycomb pattern to produce large reefs that can exceed 6' (2 m) across. The distinctive black bristles are visible only when viewed underwater, such as in a tidepool. This species also has an operculum (trap door) to close the end of its tube. When water swirls over the worm, it opens its operculum and uses its tentacles to capture plankton, organic detritus and sand grains. The food is transferred to its mouth while the sand is transported to a special organ, which coats the sand with cement used for tube building.

Individual worms are found occasionally. This one is out of its tube.

California Ice Cream Cone Worm
Pectinaria californiensis

OTHER NAMES: Ice cream cone worm; formerly *P. belgica*.
DESCRIPTION: Body is pink overall with a conical shape, flattened head and 2 sets of 14 long golden bristles. The worm makes its remarkable **straight, cone-shaped tube** by cementing fine, red-brown grains of sand together.
SIZE: To 2 1/2" (6 cm) long and 3/8" (1 cm) wide.
HABITAT: On sand and mud bottoms from the very low intertidal zone to subtidal waters.
RANGE: Alaska to Baja California, México.
NOTES: This worm positions itself in the sand, usually with its narrow end up, and uses its beautiful golden bristles to dig so that its feeding tentacles can find food lower down in the sandy or muddy substrate. The worm's cone-shaped body fits tightly inside its distinctive cone. A similar species, the trumpet worm *Pectinaria granulata*, is easily identified with its curved cone.

Coarse-tubed Pink Spaghetti Worm
Thelepus crispus

OTHER NAMES: Shell binder worm, hairy-headed terebellid worm, hairy gilled worm.
DESCRIPTION: Body is pinkish with 3 pairs of bright red gills and many elongated feeding tentacles at its anterior end. Its tube is made up of coarse sand and other objects.
SIZE: To 11" (28 cm) long.
HABITAT: Under rocks from the mid- to low intertidal zone.
RANGE: Alaska to southern California; India.
NOTES: This worm builds a **distinctive tube comprised of sand, shell fragments, small stones** and similar objects. It then resides inside its permanent burrow, extending its spaghetti-like feeding tentacles to the surface. The fine-tubed pink spaghetti worm *Thelepus setosus*, a similar looking species, builds its permanent burrow using only fine grains of sand.

MARINE WORMS

Brown Intertidal Spaghetti Worm
Eupolymnia heterobranchia

OTHER NAME: Formerly *Lanice heterobranchia*.
DESCRIPTION: Color of body varies from dark brown to greenish brown. Many tentacles extend from the anterior end. This species resides inside a parchment-like tube covered with sand granules and larger objects.
SIZE: To 4" (10 cm) long.
HABITAT: In mixtures of sand, pebbles and mud in the intertidal zone.
RANGE: Alaska to México.
NOTES: This worm feeds with its tentacles, each of which is grooved and covered with tiny cilia to bring food to the mouth. Foods include detritus, diatoms, dead and dying crustaceans, and both living and dead segmented worms.

Spiny Christmas Tree Worm
Spirobranchus spinosus

OTHER NAMES: Spiny spiral-gilled tube worm; *Spirobranchia spinosus*.
DESCRIPTION: Plume is commonly red but can be any of a wide variety of colors. The plume spirals in 3 concentric whorls.
SIZE: Plume to 3/4" (1.8 cm) in diameter. Worm to 1" (2.5 cm) long and 1/8" (3 mm) in diameter.
HABITAT: Tubes are attached to rocks from the intertidal zone to water 40' (12 m) deep.
RANGE: Central California to Baja California, México.
NOTES: The beautifully intricate detail of this species' plume is complemented by its variable colors. The plume can be appreciated only with patience as it is retracted upon the slightest of disturbances. The worm is also found burrowed into coralline algae.

Shell-Binding Colonial Worm
Chone ecaudata

Other Names: Formerly *C. minuta*, *Jasminiera ecaudata*.
Description: Outer color of the colony varies with the shell fragments or coarse sand that covers them. The plume is tan in color.
Size: Worm to 1/2" (1.3 cm) long.
Habitat: On rock adjacent to sandy areas in the intertidal zone.
Range: Vancouver Island, BC, to California.
Notes: The plume of this worm, which takes up a quarter of its length, is only visible while the worm feeds underwater. This species lives in a tube with a colony of hundreds of other worms. From shore, a colony of sand-binding colonial worms somewhat resembles a group of aggregating anemones (see p. 29) out of water. Colonies of this species are found among algae or surf-grass where reef-like colonies are formed.

Close-up of worms feeding.

- -

Spiral Tube Worm
Spirorbis sp. and others

Other Name: Tiny tube worm.
Description: Small white snail-like shell.
Size: To 1/4" (6 mm) in diameter.
Habitat: Attached to rocks, shells and various seaweeds, low intertidal zone to subtidal depths.
Range: Pacific coast.
Notes: The spiral tube worm produces a hard, snail-like shell, which is attached to a rock or similar substrate. A few similar species may be found, even on the same rock. Some individuals have a right-handed (dextral) spiral and others a left-handed (sinistral) spiral. One species is dextral in most northern populations from Oregon north; sinistral spirals are more common from California south.

MARINE WORMS

CALCAREOUS TUBE WORM
Serpula columbiana

OTHER NAMES: Red tube worm; mistakenly *Serpula vermicularis*.
DESCRIPTION: White tube is graced with many branched tentacles arranged in 2 spirals. Color of tentacles ranges widely from red to orange, pink and other colors, all with white banding.
SIZE: To 2 1/2" (6.5 cm) long.
HABITAT: On rocks in tidepools and stones in sheltered and exposed situations, low intertidal zone to water 330' (100 m) deep.
RANGE: Alaska to northern California.
NOTES: In 1767 the calcareous tube worm was first named by Carolus Linnaeus, the creator of the system by which we still classify all plants and animals. This very common worm has a visible red operculum (trap door) when its tentacles are retracted in the tube. The frilly circular tentacles filter tiny microorganisms from the water but disappear instantly with the slightest disturbance.

PEANUT WORMS
Phylum Sipuncula

The peanut worm has two parts, one of which is much larger and more globular than the other. Only 320 species in this small group have been identified worldwide.

BUSHY-HEADED PEANUT WORM
Themiste pyroides

OTHER NAMES: Flowering peanut worm, tan peanut worm, burrowing peanut worm, common peanut worm; formerly *Dendrostomum pyroides, D. petraeum, Dentrostoma patraeum.*

DESCRIPTION: This is a plump, medium brown worm with a light brown neck-like extendable portion. A tip of **bushy tentacles** rises from stems on the feeding end of the worm. Many **small brown spines are found along the "neck"** region.

SIZE: To 8" (20 cm) long and 2" (5 cm) wide.

HABITAT: Under rocks and in crevices from the low intertidal zone to shallow subtidal waters.

RANGE: Vancouver Island, BC, to Baja California, México.

NOTES: This worm extends its tentacles while underwater in order to feed on minute organic particles while under the protection of a rock or similar situation. These tentacles are highly branched and covered with mucus, which aids the worm in collecting food. This species is also known to inhabit abandoned burrows made by rock-boring clams.

AGASSIZ'S PEANUT WORM
Phascolosoma agassizii

OTHER NAME: Peanut worm.

DESCRIPTION: Light to dark brown, occasionally with purple or brown spots. Narrow, extendable **neck-like portion of body has dark bands**, followed by a wider, rough body section.

SIZE: To 4¾" (12 cm) long, ½" (1.3 cm) wide; often much smaller.

HABITAT: In sand under rocks or among the root-like holdfasts of seaweeds, above low intertidal zone.

RANGE: Alaska to Baja California, México.

NOTES: This species is the most commonly encountered peanut worm living along the Pacific coast. A few short tentacles, located near the mouth, are used to feed on detritus when the worm leaves its resting spot. This species is also known to live in burrows abandoned by hole borers such as the rough piddock (see p. 156).

MARINE WORMS

ECHIURAN WORMS
Phylum Echiura

The echiuran worms or spoonworms are a small group that are closely related to segmented worms, but they lack segments. The worm has a non-retractable proboscis (snout) that contains its brain. Echiura means "spiny tail" in Greek, a reference to ring(s) of bristles that circle the end of the worm.

FAT INNKEEPER WORM
Urechis caupo

OTHER NAME: Innkeeper worm.
DESCRIPTION: Color pinkish to yellowish pink. Resembles a sausage with 2 hook-like setae for digging near the mouth and 10–11 setae in a circle at the rear.
SIZE: To $7^{1/4}$" (18 cm) and occasionally to 20" (50 cm) long.
HABITAT: On sandy mudflats from the low intertidal zone to shallow subtidal waters.
RANGE: Humboldt Bay to Tijuana Slough, San Diego County, California.
NOTES: The fat innkeeper makes a U-shaped tunnel in sandy mudflats to a depth of 18" (45 cm). A hood-like collar is often present at the entrance to its burrow. Water is circulated though this tunnel by rhythmic contractions of the body. A filter feeder, this worm spins a fine mucus net to capture food particles. This fascinating worm got its name from the fact that several other species often coexist with it inside its tunnel. Permanent tenants here include the scaleworm *Hesperonoe advertor* and two small crabs, *Scleroplax granulata* and *Pinnixa franciscana*. Temporary tenants include the goby *Clevelandia ios*, which uses the worm's burrow to avoid enemies and to keep from drying up at low tide.

Pismo Clam *Tivela stultorum*.

MOLLUSKS AND LAMPSHELLS
Phyla Mollusca and Brachiopoda

MOLLUSKS AND LAMPSHELLS

The mollusks are a large group (phylum) of creatures, including the chitons, abalone, limpets, snails, nudibranchs, clams, mussels and octopus. Mollusks have inhabited salt water, fresh water, land and even the air for short distances. They are highly diverse, having only a few characteristics in common. All mollusks possess a fold of soft flesh (mantle) which encloses several glands, such as the stomach and the shell-producing glands. Most mollusks also have a toothed or rasping tongue (radula) and a shell covering. Scientists estimate there are some 50,000 to 130,000 species of mollusks in the world.

CHITONS
Class Polyplacophora

Chitons, sometimes referred to as sea cradles and coat-of-mail shells, range in color from bright to well camouflaged. They have a series of 8 plates or valves held together by an outer girdle. Individuals in this group can be difficult to identify as they are very similar in appearance.

LINED CHITON
Tonicella lineata

OTHER NAME: Lined red chiton.
DESCRIPTION: Striking colors varying from pink to orange-red. This chiton is named for the alternating light and dark zigzag lines on the plates. Lines are shaped in a **classic arch on the first or head valve**. Outer girdle is dark, often with light blotches.
SIZE: To 2" (5 cm) long, often much shorter.
HABITAT: On rocks with encrusting coralline algae, low intertidal to shallow subtidal zone.
RANGE: Alaska to Channel Islands, California.
NOTES: This is one of the most beautiful chitons found in California, with brilliant colors that are not easily forgotten. Its color closely matches the pink coralline algae on which it is most commonly found feeding. This chiton's main enemies are the purple sea star (see p. 197) and the six-rayed star (p. 195).

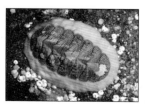

LOKI'S CHITON
Tonicella lokii

OTHER NAME: Formerly included with *Tonicella lineata*.
DESCRIPTION: Color of valves varies from salmon to light orange with vivid blue and dark maroon-brown or black zigzag pattern. **A zigzag pattern is present on the front or head valve that lacks brown lines.** Orange, pink or maroon triangles are found along the central ridge (jugal or jugum area).
SIZE: To 2" (5 cm) long.
HABITAT: On rocks covered with encrusting coralline algae from the low intertidal zone to water 79' (24 m) deep.
RANGE: Shelter Cove, Humboldt County, California, to San Miguel Island, California.
NOTES: The scientific name of this species comes from Loki, the Norse god of mischief and deception—very appropriate, as the chiton's identity eluded scientists until 1999. That year Roger Clark, a chiton specialist, recognized Loki's chiton as being new to science. It had long been believed that this species was merely a variation of the lined chiton (see p. 58).

BLUE-LINE CHITON
Tonicella undocaerulea

OTHER NAME: Formerly included with *Tonicella lineata*.
DESCRIPTION: Color of valves varies from pink to light orange with **blue zigzag lines without dark brown or black highlights**. Short dark maroon streaks often extend into the pleural areas (triangular area adjacent to the dorsal ridge).
SIZE: To 1 1/2" (3.8 cm) long.
HABITAT: On boulders and rock covered with encrusting coralline algae (see p. 240) in the low intertidal zone to waters 125' (38 m) deep.
RANGE: Kodiak Island, Alaska, to San Miguel Island, California.
NOTES: This chiton is very similar to but smaller than the lined chiton (see p. 58). An important identifying feature is a **zigzag pattern on the head valve**. Appropriately enough, part of this chiton's scientific name is derived from *caeruleus* ("sky blue"). This species is also found in Russia and Japan.

GOULD'S BABY CHITON

Lepidochiton dentiens

OTHER NAME: Formerly *Cyanoplax dentiens*.
DESCRIPTION: Somewhat elongated shape, **convex side slopes**, dark girdle with light mottling or white spots. Color of plates varies greatly, often dark brown or green. **The outer edges of the animal are nearly parallel**.
SIZE: To 1/2" (1.3 cm) long.
HABITAT: Low to mid-intertidal zone, occasionally higher in tidepools.
RANGE: Hinchinbrook Island, Alaska, to Punta Santo Tomás, Baja California, México.

NOTES: This is a common species north of Point Conception, but because of its size and coloration it is often missed. True to its common name, this is truly a small species—much smaller than most other chitons found intertidally.

KEEP'S CHITON

Lepidochitona keepiana

DESCRIPTION: Color is light overall and varies greatly from gray to green, white, yellow, orange or brown, as well as often being mottled. It is oval in shape with straight side slopes. **The outer edges of the animal are not parallel** and the girdle is narrow.
SIZE: To 5/8" (1.6 cm) long.
HABITAT: In tidepools from the mid-intertidal zone to subtidal waters 33' (10 m) deep, where it is found under small rocks.
RANGE: Otter's Point, Monterey Bay, California, to Laguna San Ignacio, Baja California Sur, México.
NOTES: Keep's chiton is a small species that is common in areas protected from waves. It prefers areas in which water stays on the shore as the tide recedes. This species is very similar to Gould's baby chiton (see above). Unfortunately little is known regarding its natural history.

Hartweg's Chiton
Lepidochitona hartwegii
OTHER NAMES: Formerly *Cyanoplax hartwegii, Lepidochiton hartwegii.*
DESCRIPTION: Valves are olive green, often with brown stripes; girdle is banded or mottled. The oval body is **low in profile** with a narrow outer girdle. **Valve surface appears very smooth to the unaided eye** although it is microscopically granulated.
SIZE: To 1 3/4" (4.5 cm) long.
HABITAT: Attached to rocks under algae in the mid-intertidal zone and in high tidepools.
RANGE: Battle Rock, Port Orford, Oregon, to Punta Abreojos, Baja California Sur, México.
NOTES: Hartweg's chiton inhabits areas protected from strong surf. It is easily found, sometimes in concentrations exceeding 25 individuals per square yard, while it rests under and feeds on spindle-shaped rockweed (see p. 236), one of its main food items. A variety of other algae make up the remainder of its diet. This chiton forages by night and often returns to the same home spot to rest for several days in a row. Researchers have discovered that Hartweg's chiton can be a simultaneous hermaphrodite. Small "males" found in the fall eventually change to "females" in the spring; these individuals can fertilize their own eggs.

Merten's Chiton
Lepidozona mertensii
DESCRIPTION: Valves range from brown to brick-red or purple in color, with intermittent white lines, giving this species a mottled look. Tiny knob-like projections give it a sandpapery appearance.
SIZE: To 2" (5 cm) long.
HABITAT: Under rocks, in low intertidal zone to water 300' (90 m) deep.
RANGE: Cook Inlet, Alaska, to northern Baja California, México.
NOTES: This is a common species, especially in the northern part of its range. Its colors help to identify it. As in all chitons, light-sensitive organs called aesthetes are found on its plates. These specialized organs help chitons retreat from light, since they do not have eyes.

COOPER'S CHITON
Lepidozona cooperi

OTHER NAME: Formerly *Ischnochiton cooperi*.
DESCRIPTION: Color ranges from dull gray to olive or brown. Plates have raised portions on the sides.
SIZE: To 1 1/2" (4 cm) long.
HABITAT: Under rocks, in intertidal zone to water 65' (20 m) deep on the open coast.
RANGE: Neah Bay, Washington, to Baja California, México.
NOTES: This chiton, like most of its relatives, is usually found under rocks away from sunlight, which ensures that it will not be easily seen by predators and will not dry out in the heat of the sun.

SMOOTH LEPIDOZONA
Lepidozona intersticta

OTHER NAMES: Formerly *Ischnochiton interstinctus*; includes *I. radians*.
DESCRIPTION: Color varies widely from white to orange or black and often spotted. Valve surface is sculptured slightly with radiating rows. **Valves appear smooth to the unaided eye**, although they are minutely granulated. Central valves are not noticeably raised and lack radiating grooves. Body is oval with a wide girdle.
SIZE: To 1" (2.5 cm) long.
HABITAT: Under rocks from the low intertidal zone to water 120' (36 m) deep.
RANGE: Prince William Sound and the Aleutian Islands, Alaska, to San Pedro, Los Angeles County, California.
NOTES: In central California, the smooth lepidozona is commonly found in the intertidal zone. In Alaska and southern California, however, it is only found in subtidal situations. This species is often brightly colored. California populations spawn in February.

Pectinate Lepidozona

Lepidozona pectinulata

OTHER NAMES: Scaled chiton, trellised chiton; formerly *L. californiensis, Ischnochiton californiensis.*
DESCRIPTION: Color of the valves and girdle varies from brown to green, yellow or orange. Varying numbers of **beaded ribs are present on the valves**. There are 20 ribs on the head valve and 5–6 on the lateral areas. **Large overlapping scales cover the girdle.**
SIZE: To 1 1/2" (4 cm) long.
HABITAT: Under rocks from the mid- to low intertidal zone.
RANGE: Cayucos, San Luis Obispo County, California, to Punta Abreojos, Baja California Sur, México.
NOTES: The pectinate lepidozona spawns in December. Little is known about the natural history of this species.

- - -

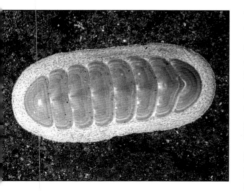

Regular Chiton

Lepidozona regularis

OTHER NAMES: Butterfly chiton, slaty blue chiton, *Ischnochiton regularis.*
DESCRIPTION: Color of both girdle and valves is an even **slate blue to olive green**. The girdle is scaled and narrow. The surface of the valves appears **smooth** to the unaided eye. The body shape is oblong overall.
SIZE: To 2" (5 cm) long.
HABITAT: Under rocks and between crevices in the low intertidal zone.
RANGE: Union Landing, Mendocino County, California, to Punta Gorda, Monterey County, California.
NOTES: This is a beautiful species, whose distinctive color and smooth surface greatly aid in identifying it. The regular chiton prefers protected areas where there are no strong waves. It spawns in February.

CALIFORNIA SPINY CHITON
Nuttallina californica

OTHER NAMES: California chiton, Nuttall's chiton, California Nuttall chiton; incorrectly thought by some to be a synonym of *N. fluxa*.

DESCRIPTION: The valves are black or dark brown, occasionally with a white stripe down each side of the dorsal ridge. Girdle is cream colored and fleshy. The girdle has **short, primarily brown bristles**; a few white bristles may be present. The length of each valve is approximately equal to the width. Gills extend almost the entire length of the foot. **Body length is $2^{1/2}$ to 3 times the width.**

SIZE: To 2" (5.1 cm) long and $^{5/8}$" (1.6 cm) wide.

HABITAT: On rocks in areas of high surf from the high to mid-intertidal zone.

RANGE: Sonoma County, California, to Punta Santo Tomás, Baja California, México.

NOTES: The California spiny chiton rests in a depression, which it uses as its home scar. When the tide returns, this common species leaves its rest area to feed on calcareous algae, including *Corallina* sp. (see p. 241), as well as several other red and green algae. These algae are often found growing on the chiton's valves. Gulls appear to be this species' main enemy.

Southern Spiny Chiton
Nuttallina scabra

OTHER NAMES: Troglodyte chiton; formerly *N. fluxa*.
DESCRIPTION: Valve color varies from green to brown or black with a brown girdle. Girdle is banded with light and dark patches. Body is elongated with a smooth central ridge and cross ribbing. **The length of each valve is considerably shorter than the width.** Valves with radiating ridges are visible when not eroded. **Girdle** is wide **with many thick, white bristles.** Gills extend 75–80% the length of the foot. **Body length is about 2 times the width.**
SIZE: To 1" (2.5 cm) long.
HABITAT: On rocks in exposed sites from the high to mid-intertidal zone.
RANGE: Monterey County, California, to Isla Asunción, Baja California Sur, México.

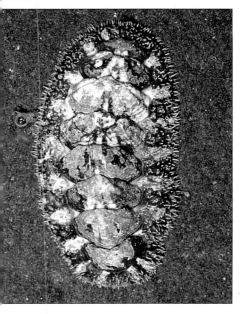

NOTES: The southern spiny chiton is a very common and colorful species favoring areas of strong surf. It is often found resting in the eroded pits of rocks. Unlike most chitons, it does not retreat from sunlight during the day. This species lives 20 years or longer.

Another similar species, the cryptic spiny chiton, *Nuttallina* sp. Nov., is soon to be described by scientists. This colorful species resembles the southern spiny chiton but its valve or shell length is approximately equal to the width. The gills are only about 75% as long as the foot.

MOLLUSKS AND LAMPSHELLS

CONSPICUOUS CHITON
Stenoplax conspicua

OTHER NAMES: Formerly *Ischnochiton conspicuus, Stenoplax sarcosa*.
DESCRIPTION: Valves green to gray or brown, occasionally with mottling. Older eroded specimens are often tinged with pink. Girdle brownish to green with a medium width and covered with dense short bristles, giving it a velvety look. **Body length is twice its width.** Valves have strong radiating ribs, **surface of the valves appears smooth to the unaided eye**.
SIZE: To 4" (10 cm) long.
HABITAT: On the undersides of rocks partially buried in sand in the low intertidal zone.
RANGE: Carpinteria, Santa Barbara County, California, to Isla Cedros, Baja California, México, and Gulf of California, México.
NOTES: Contrary to its name, the conspicuous chiton is not conspicuous until the rock it is attached to is removed from its resting spot. This striking species then becomes conspicuous with its large size and light coloration. Several individuals are often found living together attached to the underside of the same rock. This chiton is also occasionally found in holes created by the purple sea urchin (see p. 203). It is occasionally preyed upon by various octopus species.

GEM CHITON
Chaetopleura gemma

OTHER NAME: Formerly *Ischnochiton marmortus*.
DESCRIPTION: Color of valves and girdle varies from olive to red or yellow overall. Some populations are orange to brick red with a black spot on the tail valve while others are green or brown with a similar tail valve. The girdle is narrow and leathery.
SIZE: To $3/4$" (1.8 cm) long.
HABITAT: On rocks from the mid- to low intertidal zone.
RANGE: Vancouver I., BC, to Bahía Magdalena, Baja California Sur, México.
NOTES: The gem chiton is a tiny species with distinctive coloration. In California, this chiton spawns in June and has been found to be abundant in the Monterey Bay area. Short transparent spicules (small spines) are found on the girdle, but these are difficult to see.

Heath's Chiton
Stenoplax heathiana

OTHER NAME: Formerly *Ischnochiton heathiana*.
DESCRIPTION: Color of shells or valves is cream mottled with browns, lavenders and greens. Girdle is buff to brown in color with darker mottling. **Body length is twice the width.** The girdle and the surface of the valves have **scales that appear somewhat granular to the naked eye.**
SIZE: To 3" (7.5 cm) long.
HABITAT: On the undersides of rocks partially buried in sand, from the mid- to low intertidal zone.
RANGE: Union Landing, Mendocino County, California, to Punta Santo Tomás, Baja California, México.
NOTES: Heath's chiton is a nocturnal species that may be observed early in the morning before it returns to its daytime resting spot. This large chiton lays its eggs in a unique way: attached together in two elongated coils that are enclosed in a jelly-like substance. Each coil contains an average of 116,000 eggs. Most chitons lay their eggs one by one.

Hairy Chiton
Mopalia ciliata

OTHER NAME: *M. wosnessenskii*.
DESCRIPTION: Valves vary in color and are often very colorful. Wide outer girdle, covered in **soft hairs**, with a distinct notch at the rear.
SIZE: To 3" (7.5 cm) long.
HABITAT: On protected sites, such as under rocks, mid- to low intertidal zone.
RANGE: Alaska to Baja California, México.
NOTES: This chiton feeds at night and on cloudy days, grazing on tiny animals and diatoms that it finds attached to rock. Its radula (rasping tongue) contains magnetite, a hard oxide of iron that aids the animal greatly while feeding. A similar species, the mossy chiton (see p. 68), can be distinguished by the stiff hairs on its girdle. A gentle touch will determine its texture.

MOLLUSKS AND LAMPSHELLS

MOSSY CHITON
Mopalia muscosa

DESCRIPTION: Valves or dorsal plates are brown, gray or black, occasionally with white stripes. Girdle is covered in **stiff hairs**, making it look somewhat fuzzy. Notch present at rear.
SIZE: To 2 3/4" (7 cm) long.
HABITAT: Often on top of rocks and in tidepools, intertidal zone.
RANGE: Alaska to Baja California.
NOTES: This species is often observed in daylight since it does not hide under rocks as most chitons do. It stays in one place until darkness falls, when it begins feeding on algae. The mossy chiton can be distinguished from the hairy chiton (p. 67) by gently touching its girdle, which has stiff hairs. Individuals have a home range of 20" (50 cm) in the tidepool that forms their permanent home. This chiton is often found with a variety of other intertidal life forms growing on its back. Accumulations of silt do not affect it.

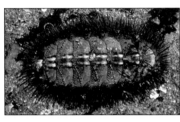

WOODY CHITON
Mopalia lignosa

DESCRIPTION: Valves can vary widely in color from brown, blue or green to gray, often with additional stripes in brown. **Stiff hairs stem from light-colored spots on girdle.**
SIZE: To 2 3/4" (7 cm) long.
HABITAT: From the mid-intertidal zone to subtidal waters.
RANGE: Prince William Sound, Alaska, to Point Conception, Santa Barbara County, California.
NOTES: This common species is often found under or on the sides of large rocks. It feeds on a variety of food, including diatoms and more than two dozen species of other algae, chiefly sea lettuce (see p. 223). The woody chiton has been observed reproducing in captivity. Females release their eggs in single file, and the eggs bunch up behind them. Males release their sperm into the water in spurts, fertilizing the eggs.

Hind's Mopalia
Mopalia hindsii

Other Name: Encrusted hairy chiton.
Description: Valves are brown to olive in color. The wide girdle contains fine hairs, distributed throughout. In some individuals a white spot is found at the top of each valve. **Rows of fine beaded sculpture cover the valves.** A prominent cleft is present at the posterior end of the girdle. Slender hairs are found scattered over the girdle.
Size: To 4" (10 cm) long.
Habitat: On rocks and pilings from the mid- to low intertidal zone.
Range: Auke Bay, Alaska, to Ventura County, California.
Notes: This large species is commonly found in both sheltered bays and estuaries, as well as on the open coast. Unlike most other chitons, Hind's mopalia is able to live in areas with high concentrations of silt. It feeds on bryozoans, filamentous algae, amphipods and barnacles.

• •

Giant Pacific Chiton
Cryptochiton stelleri

Other Names: Gumboot chiton, giant chiton, giant red chiton.
Description: Red-brown girdle completely covers plates on dorsal side.
Size: To 13" (33 cm) long.
Habitat: Low intertidal zone to subtidal waters 65' (20 m) deep.
Range: Aleutian Islands, Alaska, to San Miguel Island and San Nicolas Island, Channel Islands, California.
Notes: This species is often called the gumboot chiton, probably because of its rubbery appearance. It feeds on red algae and is known to live longer than 20 years. Small individuals were once considered edible by coastal aboriginal people. The giant Pacific chiton hosts a worm (red-banded commensal scale worm, see p. 48) that can live in the grooves on the underside of the chiton's body. The lurid rocksnail (see p. 101), which grows only to 1 1/2" (4 cm) long, has been known to attack this chiton, acclaimed as the largest chiton in the world. But the snail merely eats a shallow pit in the chiton's back.

BLACK KATY CHITON
Katharina tunicata

OTHER NAMES: Leather chiton, black chiton.

DESCRIPTION: A brown to black girdle covers most of this chiton. A white diamond shape is left uncovered on the top of each valve.

SIZE: To 4 3/4" (12.1 cm) long.

HABITAT: Commonly associated with exposed, rocky shorelines, mid-intertidal zone.

RANGE: Aleutian Islands, Alaska, to Point Conception, Santa Barbara County, California.

NOTES: This species is often found exposed during the day, feeding on algae growing on wave-washed rocks. It is large enough to have been used as food by coastal aboriginal people at one time. This chiton has a life span of only 3 years.

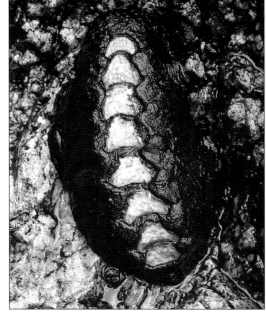

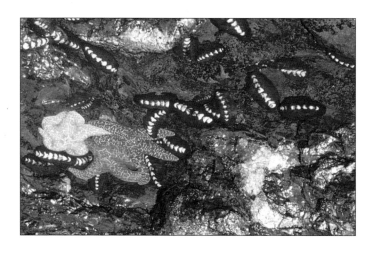

GASTROPODS (ABALONE, LIMPETS, SNAILS)
Class Gastropoda

The gastropods are a diverse group of invertebrates with few features in common, besides the muscular "foot" running along the underside of the body for locomotion (*gastropod* means "stomach foot"). Grazers, herbivores, scavengers and predators of many kinds have a specialized radula (tooth-bearing tongue) for feeding. Another specialized organ—the otocyst, similar to our hearing apparatus—can also be found in the foot of many mollusks, but is used to maintain balance.

BLACK ABALONE
Haliotis cracherodii

DESCRIPTION: Outer shell bluish to greenish black. Inner shell surface silvery, and usually **lacking a muscle scar**. Normally **5–8 open holes that are flush with the upper surface** of the smooth, oval shell. Tentacles and epipodium (fringe of skin circling the foot) are dark black.

SIZE: To 6" (15 cm) long and 4 1/4" (10.6 cm) wide normally, but larger individuals have been recorded.

HABITAT: On rock in the high intertidal zone and occasionally to 35' (10.5 m) deep.

RANGE: Coos Bay, Oregon, to Baja California, México.

NOTES: The black abalone feeds on a variety of seaweeds, including feather boa kelp (see p. 231). The shell of this abalone is normally free of marine growth. Enemies include various octopuses, purple star (p. 197), fishes and the sea otter. Shells of the black abalone have been found in the middens of aboriginal coastal California peoples, indicating they were a food source.

MOLLUSKS AND LAMPSHELLS

RED ABALONE
Haliotis rufescens

DESCRIPTION: Outer shell brick red but often covered with a wide variety of algae. Inner shell surface iridescent green, pink and copper with a central muscle scar. Normally **3–4 oval holes with raised edges** on the upper surface of the shell. Tentacles are dark black and epipodium (fringe of skin circling the foot) is black, sometimes with alternating gray stripes.
SIZE: To 12" (30 cm) long and 9 1/4" (23 cm) wide.
HABITAT: In rock crevices with heavy surf, from the low intertidal zone to water 600' (180 m) deep.
RANGE: Sunset Bay, Oregon, to Bahía Tortugas, Baja California Sur, México.
NOTES: The red abalone is a commodity prized by humans and by the sea otter. Other predators include a variety of sea stars, crabs and octopuses. This abalone feeds on loose algae fragments, including giant perennial kelp (see p. 234) and bull kelp (p. 233). The yellow boring sponge (p. 17) sometimes makes its home on the shell's exterior, which can drastically reduce its strength. The red abalone is known to live longer than 20 years.

GREEN ABALONE
Haliotis fulgens

OTHER NAME: Blue abalone.
DESCRIPTION: Outer shell from olive green to reddish brown. Exterior of the oval shell usually displays a series of 30–40 fine spiral ribs (threads) and **5–6 open, circular holes with slightly raised edges**. Shell interior is a mix of iridescent blue, copper and green with a large, beautifully iridescent muscle scar. **Tentacles are green or gray** and the epipodium (fringe of skin circling the foot) is brown or olive.
SIZE: Normally to 8" (20 cm) long and occasionally to 10" (25 cm).
HABITAT: On or under rocks from the low intertidal to water 60' (18 m) deep.
RANGE: Point Conception, Santa Barbara County, California, to Bahía Magdalena, Baja California Sur, México.
NOTES: The green abalone shows a preference for areas with strong waves. It feeds on algae fragments, trapping them with its foot as they drift by. Spawning takes place from early summer to early fall, when up to 3.5 million eggs are laid.

Rough Keyhole Limpet

Diodora aspera

DESCRIPTION: Color varies from light brown to gray, often with color banding. Prominent **circular hole** at apex of shell. Ridges radiate from top and concentric lines cross the ridges at right angles.
SIZE: To 2 3/4" (7 cm) long.
HABITAT: On rocky beaches, low intertidal to subtidal zones.
RANGE: Alaska to southern California.
NOTES: The rough keyhole limpet has a large number of teeth on its radula (tongue) for grazing on seaweed attached to rock. To protect itself from predators such as the purple star (see p. 197), this limpet erects a thin, soft mantle to cover its shell and prevent the star from attaching with its tube feet. The red-banded commensal scale worm (p. 48) is sometimes found on the underside of this limpet.

Volcano Keyhole Limpet

Fissurella volcano

OTHER NAMES: Keyhole limpet, volcano limpet.
DESCRIPTION: Shell exterior is pink to red with several black or reddish brown lines radiating from the top. Its foot is yellowish. There is a **small elongated opening at the apex (tip) of the shell**.
SIZE: To 1 3/8" (3.5 cm) long.
HABITAT: On rocks in the mid-intertidal zone.
RANGE: Crescent City, Del Norte County, California, to Bahía Magdalena, Baja California Sur, México.
NOTES: The volcano limpet is preyed upon by the purple star (see p. 197), from which it has been observed to "run" away. This species gets its name from the splashes of purple or red that resemble lava flowing down the slopes of a volcano. This limpet uses the opening in its shell to pass wastes, as well as water that has passed through the gills.

TWO-SPOT KEYHOLE LIMPET
Fissurellidea bimaculata

OTHER NAMES: Two-spotted keyhole limpet; formerly *Megatebennus bimaculatus*.
DESCRIPTION: The fleshy body varies widely in color from red, orange or brown to white mottled with brown. The shell has a **broadly elongated hole** that is approximately **a third of the shell length**. When the shell is placed on a flat surface the ends are noticeably upturned.
SIZE: Shell to 3/4" (2 cm) long.
HABITAT: On kelp holdfasts and under rocks from the low intertidal zone to water 100' (30 m) deep.
RANGE: Sitka, Alaska, to southern Baja California, México.
NOTES: The two-spot keyhole limpet feeds on colonial tunicates and sponges. Its large, fleshy body nearly dwarfs its tiny shell, which sits on the top of its body. Harlequin ducks are known to feed occasionally on this species.

GIANT KEYHOLE LIMPET
Megathura crenulata

OTHER NAMES: Great keyhole limpet, giant key hole limpet.
DESCRIPTION: The **large, soft body** ranges in color from black or brown to mottled gray and dwarfs the shell. Exterior of shell varies from buff to pink. Fine ridges radiate from the apex of shell, on which there is a large, elongated **central oval opening approximately 1/4 the shell length**.
SIZE: Shell to 5" (13 cm) long; body to 10" (25 cm) long.
HABITAT: On rocks in protected areas from the low intertidal to subtidal waters.
RANGE: Monterey Bay, California, to Isla Asunción, Baja California Sur, México.
NOTES: The shell of this large, impressive animal is largely hidden by its body. Its diet includes algae, tunicates and a variety of encrusting creatures. Pigment from black-bodied specimens rubs off onto your hands if you handle them. The shells of the giant keyhole limpet were once used as wampum or money by Aboriginal peoples.

WHITECAP LIMPET
Acmaea mitra

OTHER NAMES: Dunce-cap limpet, Chinaman's hat limpet.
DESCRIPTION: Shell is white or pink (see below) and somewhat cone-shaped. Shell interior bears a horseshoe-shaped muscle scar.
SIZE: To 1" (2.5 cm) high.
HABITAT: On rocky beaches, low intertidal to shallow subtidal zone.

RANGE: Aleutian Islands, Alaska, to Baja California, México.
NOTES: This limpet is often found covered in pink encrusting coralline algae (see p. 240), which is also its prime food. The limpet's foot is strong enough to keep it from being washed away in the strongest of waves along the exposed coast.

SHIELD LIMPET
Lottia pelta

OTHER NAME: *Collisella pelta*.
DESCRIPTION: Brown oval shell exterior with a variety of markings, occasionally with wavy edge. **Apex of shell is relatively high**.
SIZE: To $2^{1/8}$" (5.4 cm) long and $5/8$" (1.5 cm) high.
HABITAT: On rocks and among mussel beds, between high and low intertidal zones. Also found on various species of brown algae, including the surf-loving sea palm (see p. 232).
RANGE: Aleutian Islands, Alaska, to Bahía del Rosario, Baja California, México.
NOTES: The limpet, like many other mollusks, enlarges its own shell. A group of glands around the edge of the mantle secrete a "liquid shell," made primarily of carbonate of lime, which hardens over time. In this way, the limpet's home is always the correct size. This limpet is similar to another species, the plate limpet (see p.76), which possesses a very flat shell.

MASK LIMPET
Tectura persona

OTHER NAMES: Speckled limpet, masked limpet, *Notoacmea persona*.
DESCRIPTION: Color varies from blue-gray to brownish, with whitish rays stemming from top or a pattern of light gray spots. Several tiny white spots can be seen on top of shell. Apex is markedly off-center with a **slight hook-like shape near the tip**.
SIZE: To 1 1/2" (4 cm) long.
HABITAT: Prefers the dark of rock crevices or similar areas in high intertidal zone but comes out at night to feed.
RANGE: Alaska to Monterey, California.
NOTES: Once the sun has set and darkness prevails, this limpet is busy feeding on algae. It has been calculated that a limpet of 1 square inch (6.5 cm^2) requires a browsing area of encrusting seaweed covering 75 square inches (487 cm^2) each year to survive.

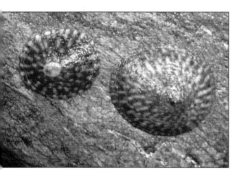

PLATE LIMPET
Tectura scutum

OTHER NAME: *Notoacmea scutum*.
DESCRIPTION: Often gray to greenish in color with off-white rectangular blotches radiating from top of the flat, oval shell. **Apex of shell is relatively low**.
SIZE: To 2" (5.1 cm) long.
HABITAT: On rocks, high to low intertidal zones.
RANGE: Alaska to Baja California, México.
NOTES: The plate limpet can be distinguished from other species by its brown tentacles, but these can be seen only if the limpet is active and not pressed tight against the substrate. This limpet tries to escape when various predatory sea stars are detected. Green algae are occasionally seen growing on its shell.

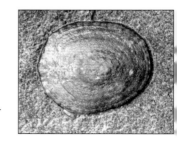

Black Limpet
Lottia asmi

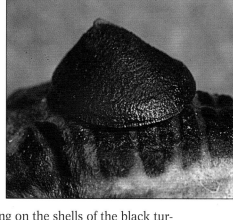

Other Names: Formerly *Collisella asmi*, *Acmaea asmi*.
Description: Shell color varies from black to dark brown and shell interior is dark. Fine radiating ridges are present on the shell's exterior. This species has a high profile.
Size: To 1/2" (1 cm) long.
Habitat: On rocks and in tidepools from the mid-intertidal zone.
Range: Northern Vancouver Island, BC, to Isla Socorro, Islas Revillagigedos, México.
Notes: The black limpet feeds on tiny algae living on the shells of the black turban (see p. 81), speckled turban (p. 83) and occasionally other snails, as well as the California mussel (p. 125). This tiny limpet moves freely from one host to another, which is especially easy when its hosts aggregate in clusters. Studies have shown that most black limpets change hosts at least once a day.

Ribbed Limpet
Lottia digitalis

Other Names: Finger limpet, fingered limpet, *Collisella digitalis*.
Description: Overall color is gray with greenish brown bands. Shell is elliptical (one end is narrower than the other); several prominent ribs radiate from top. **Edge of shell has a somewhat wavy margin**.
Size: To 1" (2.5 cm) long.
Habitat: In rocky areas, splash zone to high intertidal zone, often on vertical rocks in the shade.
Range: Alaska to Baja California, México.
Notes: One study showed that the greatest distance a ribbed limpet wandered from its established territory was only 3' (90 cm). The shape of the shell's outer edge matches precisely the rock area that is its home. This close fit is very helpful in preventing a predator from removing the limpet from its substrate.

ROUGH LIMPET
Lottia scabra

OTHER NAMES: Formerly *Acmaea scabra, Collisella scabra, Macclintockia scabra.*
DESCRIPTION: Color of shell varies from greenish to brown and is often eroded, while the fleshy portions are light and peppered with black. **Shell has strong radiating ribs, scalloped edge** and medium profile. Tentacles are white, and the foot displays lateral black spots.
SIZE: To $1^{1/8}$" (3 cm) long.
HABITAT: On horizontal or gently sloping rocks from the spray zone to mid-intertidal zone.
RANGE: Cape Arago, Oregon, to southern Baja California, México.
NOTES: The rough limpet feeds primarily on the algal film found on horizontal and angled rocks in the upper intertidal zone. The similar-looking ribbed limpet (see p. 77) also feeds on algal film, but this species usually feeds on vertical rocks in the same intertidal zone. The rough limpet is known to return to its home scar daily at each low tide. A home scar is clearly visible when its owner is not resting on it. The rough limpet is believed to live as long as 11 years.

FILE LIMPET
Lottia limatula

OTHER NAME: Formerly *Collisella limatula.*
DESCRIPTION: Shell is relatively flat. Fine, toothed ribs radiate from the off-center apex.
SIZE: To $1^{3/4}$" (4.5 cm) long.
HABITAT: Mid- to lower intertidal zone.
RANGE: Southern BC to Baja California, México.
NOTES: Occasionally albino specimens of this limpet are found. They have tan to cream-brown shell exteriors. Like most limpets, the file limpet is a vegetarian and as such has no proboscis (snout), which flesh-eating (carnivorous) mollusks do. Individuals are found as separate sexes and their young start out life as free-swimming organisms.

Giant Owl Limpet
Lottia gigantea

OTHER NAME: Owl limpet.
DESCRIPTION: Exterior is brown with whitish spots. Shell interior has distinctive owl-shaped markings. Limpet has a low profile with **apex very near the front**.
SIZE: To 3 1/2" (9 cm) long.
HABITAT: On exposed rock from the high to low intertidal zone.
RANGE: Neah Bay, Washington, to Bahía Tortugas, Baja California Sur, México.
NOTES: The giant owl limpet feeds on a variety of tiny algae growing on rock. It is a loner, known to bulldoze limpets, mussels and sea anemones off the rock within its home territory—about 1 square foot (1,000 cm^2). This territory may change during spring and summer, however, so that an individual may move as far as 50' (15 m) away. At low tide, some giant owl limpets return to their home scar, a place where they regularly return to rest and which exactly matches the shape of their shells. The giant owl limpet is one of the largest limpets found in North America.

Unstable Limpet
Lottia instabilis

OTHER NAMES: *Collisella instabilis, Acmaea instabilis.*
DESCRIPTION: Exterior dark brown, occasionally with a yellow-brown top. Interior is white with a central brown blotch. **Shell is saddle-shaped** when turned on its edge.
SIZE: To 1 3/8" (3.5 cm) long.
HABITAT: On holdfasts and stipes (stems) of various species of kelp, low intertidal zone to subtidal depths of 240' (73 m).
RANGE: Kodiak Island, Alaska, to San Diego, California.
NOTES: This limpet's distinctive shape enables it to move easily on the stipes of seaweeds. If the shell is placed on a flat surface, it can rock back and forth. The unstable limpet feeds on various seaweeds, especially the split kelp (see p. 228).

MOLLUSKS AND LAMPSHELLS

SEAWEED LIMPET
Discurria insessa

OTHER NAMES: Kelp limpet; formerly *Notoacmaea insessa, Acmaea insessa*.
DESCRIPTION: Shell exterior is light brown and glossy; interior is light brown with a white ring at edge. Apex is high near the front of the shell. The shell has many fine concentric lines.
SIZE: To 7/8" (2.2 cm) long.
HABITAT: On algae in the low intertidal zone.
RANGE: Wrangell Island, southeast Alaska, to Bahía Magdalena, Baja California Sur, México.
NOTES: This limpet is commonly found on feather boa kelp (see p. 231), where it moves freely from one frond to another and eats a depression in the central stipe (stem). Larger, older individuals orient themselves in the same direction as the stipe.

BLUE TOPSNAIL
Calliostoma ligatum

OTHER NAMES: Blue top shell, ribbed topsnail; formerly *Calliostoma costatum*.
DESCRIPTION: Beautiful shell with brown striped exterior, nearly cone-shaped with several somewhat rounded whorls. Worn shells reveal underlying blue color.
SIZE: To 1" (2.5 cm) in diameter.
HABITAT: On exposed, rocky beaches, low intertidal zone to subtidal water 100' (30 m) deep.
RANGE: Alaska to San Pedro, California.
NOTES: This snail feeds on diatoms, kelp, detritus, hydroids and bryozoans. Its enemies include the six-rayed star (see p. 195) and the purple star (p. 197), and it has been found in the stomachs of lingcod. If the presence of a predatory sea star is detected, the speed of the blue topsnail doubles to 3/32" (2.9 mm) per second.

Wavy Turban
Megastraea undosa

OTHER NAMES: Wavy top shell, wavy top snail, wavy top turban, wavy turban snail; formerly *Astraea undosa*.
DESCRIPTION: Tan in color with a heavy periostracum (shell covering). Shell is heavy with **wavy, slanted ridges on the whorls** and a concave base. Operculum (trap door) is teardrop-shaped, calcareous and heavy.
SIZE: Shell to 4 1/2" (11 cm) in diameter.
HABITAT: On rock from the low intertidal zone to water 60' (18 m) deep.
RANGE: Point Conception, Santa Barbara County, California, to Isla Asunción, Baja California Sur, México.
NOTES: The wavy turban is a large, impressive species that reaches even larger sizes at subtidal depths among the kelp. Calcareous red algae frequently grow on the shell of this slow mover. The distinctive operculum (trap door; see photo) is often found washed ashore after a storm.

The distinctive operculum of the wavy turban.

Black Turban
Tegula funebralis

OTHER NAME: Black top-shell.
DESCRIPTION: Purple-black shell with 4 rounded whorls.
SIZE: To 1 1/4" (3 cm) in diameter.
HABITAT: On rocky shores, high to mid-intertidal zone.
RANGE: Vancouver Island, BC, to Point Conception, California, with reports to Baja California, México.
NOTES: The turban eats only soft seaweed, using a radula (specialized tongue). Black turbans are believed to live as long as 100 years. Perhaps this is one reason why the tops of their shells are almost always worn to the underlying white shell layer.

BROWN TURBAN
Tegula brunnea

OTHER NAMES: Brown tegula, brown turban snail.
DESCRIPTION: Shell color varies from orange to bright brown. Shell has **smooth and rounded whorls** and one tooth on the columella (central pillar). **No umbilicus or "navel"** present, just a slight depression. Foot has dark brown to black sides and light beneath.
SIZE: To $1^{1/2}$" (3 cm) in diameter.
HABITAT: On kelp from the low intertidal zone to subtidal waters.
RANGE: Cape Arago, Oregon, to Santa Barbara Island, California.
NOTES: The brown turban is a common species occurring just below the intertidal level of the black turban (see p. 81). Brown algae are believed to be its main food. Primary predators include the purple star (p. 197) and giant pink star (p. 196).

SMOOTH BROWN TURBAN
Norrisia norrisi

OTHER NAMES: Smooth turban, smooth turban snail, Norris' top snail, Norris' top shell, Norris shell, norrissnail.
DESCRIPTION: Chestnut brown shell and green umbilicus (navel-like depression). The flattened shell is smooth with rounded whorls. Distinctive **bright red foot**.
SIZE: Shell to $2^{1/2}$" (6.5 cm) in diameter.
HABITAT: On rock from the low intertidal zone to water 100' (30 m) deep.
RANGE: Point Conception, Santa Barbara County, California, to Isla Asunción, Baja California Sur, México.
NOTES: The smooth brown turban is usually found on brown algae, especially the giant perennial kelp (see p. 234) and feather boa kelp (p. 231). Here it feeds as it moves down the algae during the day; at night it returns to higher levels.

BANDED TURBAN
Tegula eiseni

OTHER NAMES: Western banded tegula, banded tegula, beaded turban snail, *T. mendella*; formerly believed to be *Tegula ligulata*, a closely related Mexican species.
DESCRIPTION: Shell brownish with black and white spots. **Numerous knobby spiral bands** are evenly spaced on the rounded whorls. Umbilicus (navel-like depression) is open.
SIZE: To 1" (2.5 cm) in diameter.
HABITAT: On rocks and in rubble from the mid-intertidal zone to subtidal kelp forests.
RANGE: Monterey, California, to Bahía Magdalena, Baja California Sur, México.
NOTES: This common species is often found under kelp. It is reported to be nocturnal. Little is known about the biology of many turbans, including this one.

• •

SPECKLED TURBAN
Tegula gallina

OTHER NAME: Speckled tegula.
DESCRIPTION: Color of shell varies from gray to greenish with alternating slanted white and dark stripes or zigzag marks. Its whorls are convex and rough. The umbilicus ("navel") is present only as a shallow depression, if it is present at all.
SIZE: Shell to 1 3/8" (3.5 cm) in diameter.
HABITAT: On rocks and in tidepools in the mid-intertidal zone.
RANGE: Santa Barbara County, California, to Bahía Magdalena, Baja California Sur, México.
NOTES: The speckled turban is often found in the company of the black turban (see p. 81) in tidepools and among rocks. This species, like most herbivorous gastropods, leaves a trail of mucus as it moves about.

MOLLUSKS AND LAMPSHELLS

GILDED TURBAN
Tegula aureotincta

OTHER NAME: Gilded tegula.
DESCRIPTION: Shell color varies from dark gray to olive or brown. Shell spire (whorls above the large body whorl) is very low. Spiral ridges are prominent on the base. There is an **orange-yellow stain at the umbilicus** ("navel"), often surrounded by a sky blue band.
SIZE: To 1 1/2" (4 cm) in diameter.
HABITAT: On rocky shores from the mid-intertidal zone to subtidal waters.
RANGE: Ventura County, California, to Bahía Magdalena, Baja California Sur, México.
NOTES: The gilded turban is believed to be closely related to the brown turban (see p. 82), but little is known about the natural history of this species.

DARK DWARF-TURBAN
Homalopoma luridum

OTHER NAMES: Formerly *Leptothyra carpenteri, Homalopoma carpenteri*.
DESCRIPTION: Color varies greatly from gray, brown or red to white. Light bands of color may also be present. Exterior of shell is sculptured with rounded spiral ribs.
SIZE: Normally to 3/16" (5 mm) long and occasionally to 3/8" (9 mm).
HABITAT: Under rocks from the low intertidal zone to subtidal waters.
RANGE: Sitka, southeast Alaska, to Isla San Gerónimo, Baja California, México.
NOTES: This species is common but easily overlooked because of its small size. Its handsome empty shells are often used by young hermit crabs.

FLAT-BOTTOMED PERIWINKLE
Littorina keenae

OTHER NAMES: Eroded periwinkle, gray periwinkle, gray littorine; formerly *Littorina planaxis*.
DESCRIPTION: Gray-brown shell overall with a **flattened white inner lip** and columella (central pillar). Pale white spots are sometimes present. The somewhat rounded **shell is often eroded**.
SIZE: To 3/4" (1.9 cm) high.
HABITAT: On rocks from splash zone to high intertidal zone.
RANGE: Charleston, Oregon, to Bahía Magdalena, Baja California Sur, México.
NOTES: Like all periwinkles, this species is out of the water most of the time and has been known to survive out of water for up to 3 months. It is a common snail that secretes a glue-like mucus around the aperture of its shell to help it cling to rocks when it is out of water for extended periods. This periwinkle is found higher in the intertidal zone of California than any other mollusk. It feeds on a fine layer of diatoms and small algae that cling to the rocks on which it lives.

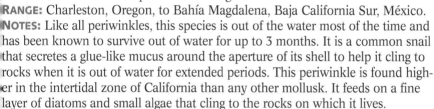

CHECKERED PERIWINKLE
Littorina scutulata

OTHER NAME: Checkered littorine.
DESCRIPTION: Smooth, brown or black **elongated shell with white checkered spots**.
SIZE: To 5/8" (1.6 cm) high.
HABITAT: On various types of seaweed or rocky shores, high and middle intertidal zones.
RANGE: Kodiak Island, Alaska, to Bahía de Tortuga, Baja California, México.
NOTES: The largest checkered periwinkles are often found much higher up on shore than smaller individuals. Their enemies include several carnivorous gastropods and the six-rayed star (see p. 195). Snails in this family have a unique foot, which is divided into two separate parts that move alternately. Some scientists believe land snails may have evolved from periwinkles.

MOLLUSKS AND LAMPSHELLS

SCALY TUBE SNAIL
Serpulorbis squamigerus

OTHER NAME: Scaled wormsnail, scaled worm snail, scaled wormshell; formerly *Aletes squamigerus*.
DESCRIPTION: Tube gray with a twisted shape and **longitudinal ribs**. Surface wrinkled and **covered with scales**.
SIZE: Shell to 5" (13 cm) long and 1/2" (1.3 cm) in diameter.
HABITAT: Attached to protected rocks and docks from the high intertidal zone to water 65' (20 m) deep.
RANGE: Monterey Bay, California, to Peru.
NOTES: The scaly tube snail is a worm-like sedentary snail that attaches to a rock or similar object. Here it secretes mucus to capture the tiny drifting particles that make up its diet. This is a gregarious species. It has been found in concentrations of up to 650 individuals per square yard.

MUDFLAT SNAIL
Batillaria cumingi

OTHER NAMES: Screw shell, Cuming's false cerith, false-cerith snail, *B. zonalis*, *B. attramentaria*.
DESCRIPTION: Tan to black with **white bands** and a bead-like finish; elongated, tapered shell with 8 or 9 **flat whorls**.
SIZE: To 1 1/4" (3.2 cm) high.
HABITAT: On mud shorelines, high to mid-intertidal zone.
RANGE: Boundary Bay, BC, to Elkhorn Slough, California.
NOTES: This species was accidentally introduced with oysters from Japan. Its common name is very appropriate, as it often lives on mud flats. The snail has been observed to occur in incredible densities—7,000 individuals per square yard in ideal habitats. It is estimated to live as long as 10 years.

California Horn Snail

Cerithidea californica

OTHER NAMES: California horn shell, California hornsnail.
DESCRIPTION: Shell color varies from yellowish brown to black. **Turret-shaped shell** with 10–12 ribbed, **well rounded whorls**.
SIZE: To 1 3/4" (4.5 cm) high.
HABITAT: On high intertidal mud flats.
RANGE: Bolinas Bay, Marin County, California, to Laguna San Ignacio, Baja California Sur, México.
NOTES: This snail is an abundant inhabitant of southern California's estuaries and lagoon mud flats, where it feeds on minute organic detritus. In these oxygen-poor muddy sites, the snail thrives in densities greater than 20 individuals per square foot.

Tinted Wentletrap

Nitidiscala tincta

OTHER NAMES: Painted wentletrap, *Epitonium tinctum*.
DESCRIPTION: The white shell may have a **brown or purplish stripe in the suture** (groove between 2 whorls). Whorls are smooth and convex with 9–13 axial ribs.
SIZE: Shell to 5/8" (1.6 cm) high for intertidal individuals and 1 1/4" (3.2 cm) high for subtidal specimens.
HABITAT: Among rocks or in tidepools from the low intertidal zone to water 150' (46 m) deep.
RANGE: Forrester Island, Alaska, to Bahía Magdalena, Baja California Sur, México.
NOTES: The tinted wentletrap lives near its food source, aggregating anemones (see p. 29) and giant green anemones (p. 30). It feeds on the tentacles of these and other anemones when the tide covers them. When the tide is out, it hides in the sand. It avoids the plumose anemone (p. 34), however, because it occasionally falls prey to it. If you handle a live tinted wentletrap, it may irritate your skin. A purple dye may also be secreted if the animal is disturbed.

BOREAL WENTLETRAP

Opalia borealis

OTHER NAMES: Chace's wentletrap, *Opalia chacei*; some consider this species a subspecies of *O. wroblewskyi*.
DESCRIPTION: Color white to cream. Axial ribs are broadly rounded with about every third rib larger and running over the keel onto the base.
SIZE: Shell to 1 1/4" (3.2 cm) long.
HABITAT: Under rocks from the low intertidal zone to subtidal waters.
RANGE: Queen Charlotte Strait, BC, to Santa Catalina Island, California.
NOTES: The boreal wentletrap is usually found in association with anemones; scientists assume that it preys on anemones as do other species of wentletrap. Little is known about its biology.

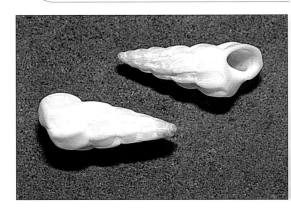

SCALLOP-EDGED WENTLETRAP

Opalia funiculata

OTHER NAMES: Scalloped wentletrap, sculptured wentletrap; formerly *O. insculpta*, *O. crenimarginata*.
DESCRIPTION: Shell varies from white to golden yellow. Whorls are smooth and slightly convex with 14 wave-like ribs giving a **scalloped edge**.
SIZE: To 3/4" (2 cm) high.
HABITAT: Near sea anemones in the low intertidal zone.
RANGE: Santa Monica, Los Angeles County, California, to Peru.
NOTES: The scallop-edged wentletrap is a carnivore that can often be found at the base of a giant green anemone (see p. 30), which it feeds on. Empty shells of this species are often inhabited by small hermit crabs. *Wentletrap* means "winding staircase," an accurate description of this distinctive group of snails.

FLAT HOOFSNAIL
Hipponix cranioides

OTHER NAMES: Ancient hoof shell, white hoofsnail, horse's hoofsnail, Washington hoof shell, *H. antiquatus, Antisabia cranioides*.
DESCRIPTION: Shell is white with a brownish periostracum (covering). It is flat and circular, and lacks any internal calcareous structures. **Muscle scar inside shell is horseshoe-shaped**. Apex is blunt at one end.
SIZE: Shell to 3/4" (2 cm) in diameter.
HABITAT: On rocks and shells from the low intertidal zone to subtidal waters on the exposed coast.
RANGE: Vancouver Island, BC, to Monterey, California, and possibly farther south.

NOTES: The flat hoofsnail is found on both Atlantic and Pacific coasts. This species secretes and sits on a calcareous slab or shell base that is not attached to but perfectly conforms to its upper shell. The animal feeds on detritus and coralline algae fragments by extending its long proboscis (snout). Its skin-like periostracum (shell covering) is usually rubbed off individuals that are washed up on the beach.

HOOKED SLIPPERSNAIL
Crepidula adunca

DESCRIPTION: Dark brown shell beneath a yellow skin-like periostracum (covering). A beak-like hook at the tip overhangs posterior edge of shell.
SIZE: To 1" (2.5 cm) high.
HABITAT: On rocks or on shells of other snails, intertidal zone.
RANGE: Queen Charlotte Islands, BC, to Punta Santo Tomás, Baja California, México.
NOTES: The hooked slippersnail is found riding the shells of various snails, including the black turban (see p. 81), which also may be used by hermit crabs. Slippersnails have both male and female parts but only become one sex at a time. Large slippersnails are females waiting for smaller male suitors. They prepare several small egg capsules, placing approximately 250 eggs in each. The capsules are attached to the surface on which the female resides and are guarded for approximately one month until the eggs hatch. The smaller slippersnails are males, which will become females and lay eggs as they grow older.

PACIFIC HALF-SLIPPERSNAIL

Crepipatella dorsata

OTHER NAMES: Pacific half-slipper, half-slipper snail, half-slipper shell, wrinkled slipper-shell, *C. lingulata*.

DESCRIPTION: Shell is white with brown mottling or radiating stripes and a yellow skin-like periostracum (covering). Shell is wrinkled, low in profile and round with an irregular margin. An interior **shelf is attached only to one side** of the shell.

SIZE: To 1" (2.5 cm) long.

HABITAT: On rocks and the shells of snails from the low intertidal zone to water 300' (90 m) deep.

RANGE: Bering Sea, Alaska, to Peru.

NOTES: This species is a filter feeder, consuming plankton and organic detritus suspended in the water. Empty shells are sometimes found washed ashore after a storm. Little is known about its natural history.

ONYX SLIPPERSNAIL

Crepidula onyx

OTHER NAMES: Onyx slipper snail, onyx slipper-shell; formerly *C. cerithicola, C. lirata*.

DESCRIPTION: Shell from tan to dark **brown exterior** and **white interior shelf. Apex is low, curved to one side** close to the margin but **not distinctly hooked**. Interior shelf is notched inward at the ends.

SIZE: To 2³/₄" (7 cm) long.

HABITAT: On rocks and shells from the intertidal zone to 300' (90 m).

RANGE: Monterey, California, to Peru.

NOTES: Two forms of the onyx slippersnail can be found. Broad-shelled individuals grow on large rocks; others live inside the shells of small snails, where they grow with very narrow shells. Like all slippersnails, the males are small and change to females when they grow larger. Up to 10 individuals have been observed to live stacked on top of each other. Food consists of plankton and organic debris.

NORTHERN WHITE SLIPPERSNAIL

Crepidula nummaria

OTHER NAMES: White slipper snail, white slipper shell.
DESCRIPTION: Shell interior is shiny white; exterior is white with a yellowish brown periostracum (covering). Shape is quite variable, but generally flat and broad.
SIZE: Shell to $1^{1/2}$" (4 cm) long.
HABITAT: On protected rocks and dead shells from the low intertidal zone to shallow subtidal depths.
RANGE: Alaska to Panama.
NOTES: The northern white slippersnail can be found living inside and outside large snail shells, but individuals living in these situations grow to only $5/8$" (1.6 cm) long.

Females release more than 5,000 eggs with each spawning. When the larvae settle by a functional female, they become males. However, a male will develop into a female if there are no functional females nearby.

SPINY CUP-AND-SAUCER

Crucibulum spinosum

OTHER NAMES: Spiny cup-and-saucer shell, cup-and-saucer limpet, spiny cup and saucer shell.
DESCRIPTION: Exterior varies from brown to yellowish with dark-colored rays. Circular **shell is covered with spines**; apex is often twisted. **Interior of shell resembles a complete cup and saucer**.
SIZE: Shell to $1^{1/8}$" (3 cm) in diameter.
HABITAT: On rocks and similar objects from the mid-intertidal zone to water 180' (55 m) deep.
RANGE: San Pedro, Los Angeles County, California, to Chile.
NOTES: The spiny cup-and-saucer is a filter feeder that can be seen living on a variety of mollusk shells. Smaller males are often found sitting on larger females. The light, empty shells of this gastropod are sometimes washed up on shore.

MOLLUSKS AND LAMPSHELLS

SAN DIEGO LAMELLARIA
Lamellaria diegoensis

OTHER NAME: Formerly *L. stearnsii*.
DESCRIPTION: The white ear-shaped shell is dwarfed by the large body.
SIZE: Shell to 3/4" (2 cm) in diameter.
HABITAT: Low intertidal zone to subtidal waters.
RANGE: Monterey Bay, California, to Santa Tomas, Baja California, México.

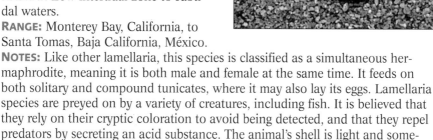

NOTES: Like other lamellaria, this species is classified as a simultaneous hermaphrodite, meaning it is both male and female at the same time. It feeds on both solitary and compound tunicates, where it may also lay its eggs. Lamellaria species are preyed on by a variety of creatures, including fish. It is believed that they rely on their cryptic coloration to avoid being detected, and that they repel predators by secreting an acid substance. The animal's shell is light and sometimes washes ashore on the beach.

APPLESEED ERATO
Hespererato vitellina

OTHER NAMES: Apple-seed erato, apple seed erato; formerly *Erato vitellina*.
DESCRIPTION: Shell is shiny overall, purplish red on the dorsal side. It is pear-shaped with a narrow opening and an outer lip with 7–10 rounded teeth.
SIZE: To 5/8" (1.6 cm) long.
HABITAT: On pilings, in gravel or on algae from the low intertidal zone to water 250' (76 m) deep.
RANGE: Bodega Bay, Sonoma County, California, to Bahía Magdalena, Baja California Sur, México.
NOTES: Empty shells of the appleseed erato are sometimes washed up on shore with beach drift. The species is aptly named: the shells closely resemble apple seeds. The snail is often found living near compound tunicates, on which it feeds using its long proboscis (snout) to pull in its food.

CALIFORNIA TRIVIA
Pusula californiana

OTHER NAMES: Little coffee-bean, little coffee bean shell, *Trivia californiana*.
DESCRIPTION: Shell is dark purple to brown overall with a total of 10–12 **white ridges radiating from a shallow central groove**. Opening is slit-like.
SIZE: To 7/16" (1 cm) long.
HABITAT: Under rocks and among seaweeds from the low intertidal zone to water 250' (76 m) deep.
RANGE: Crescent City, Del Norte County, California, to Acapulco, Guerrero, México.
NOTES: The beachcomber sometimes finds empty shells of the California trivia washed up on shore, but their small size make them a prize only for those willing to search diligently. A similar species, Solander's trivia (*Pusula solandri*) is a larger animal with a deeper central groove and a lighter color overall.

CHESTNUT COWRIE
Zonaria spadicea

OTHER NAMES: Chestnut cowry, California brown cowry; formerly *Cypraea spadicea*.
DESCRIPTION: Shell is chestnut along the middle with white to gray edges, and highly polished with an oval shape and a slit-like opening. The mantle is rosy brown speckled with black.
SIZE: Normally to 2 5/8" (6.5 cm) long but occasionally to 4 3/4" (12 cm).
HABITAT: Under ledges and in crevices from the low intertidal zone to water 165' (50 m) deep.
RANGE: Monterey, California, to Isla Cedros, Baja California, México, but rare north of Point Conception, Santa Barbara County.
NOTES: Many cowries are found in tropical and warm-temperate waters worldwide, but the chestnut cowrie is the only cowrie found in California. This beautiful species is known to dine on a wide range of foods, including sea anemones, compound tunicates, sponges, snail eggs and carrion. During the summer months, the female lays up to 100 egg capsules, each of which may contain several hundred eggs.

MOLLUSKS AND LAMPSHELLS

LEWIS'S MOONSNAIL
Euspira lewisii

OTHER NAMES: *Lunatia lewisi*; formerly *Polinices lewisii*.
DESCRIPTION: Shell is yellowish with a brown paper-like periostracum (covering). Shell is made of about 6 rounded whorls. The **umbilicus ("navel") is open, deep, round and narrow**. The tongue-like callus is often brown and rather small in size. Foot is enormous, pink and fleshy.
SIZE: To 5 1/2" (14 cm) high.
HABITAT: In sand, gravel or mud, low intertidal zone to subtidal water 165' (50 m) deep.
RANGE: Southeastern Alaska to Isla San Geronimo, Baja California, México.
NOTES: This active carnivore plows through the sand preying on clams and other bivalves by

drilling a hole in the shell of the prey. To protect itself, it contracts its foot by ejecting water from perforations around the edge. This enables the animal to retreat into the shell, filling most of it. Lewis's moonsnail is also known for its distinctive large egg case, or "sand collar." Females crawl higher up on muddy sand beaches in the spring and summer to lay their eggs. The collar is formed around the extended mantle or fleshy foot and is made of 2 layers of sand sandwiching a layer of eggs, all held together with a mucous secretion. The moonsnail moves from beneath the collar once she is finished. She then leaves the egg case on the sand for the eggs to hatch.

The egg case has a distinctive shape.

SOUTHERN MOONSNAIL
Neverita reclusiana

OTHER NAMES: Southern moon shell, Recluz's moon shell, Recluz's moon-shell, *Polinices reclusianus, P. recluzianus*.
DESCRIPTION: Shell color is tan to brown, heavy and globular with 4 or 5 whorls. The **umbilicus ("navel") is covered or nearly covered by a white, tongue-like callus**.
SIZE: Shell to 2 3/4" (7 cm) in diameter.

HABITAT: On sandy mud from the low intertidal zone to water 150' (46 m) deep.
RANGE: Crescent City, Del Norte County, California, to Mazatlan, México.
NOTES: The southern moonsnail is often seen in lagoons and shallow bays. Its shells have been found in middens, indicating that it was once a food source for Native Americans.

ANGULAR UNICORN
Acanthina spirata

OTHER NAMES: Thorn snail, spotted thorn drupe.
DESCRIPTION: **Small black markings** spiral on whorls. There is a **prominent ridge or keel at the shoulder** of the shell. Spire (upper part) of the shell is elongated. A stout spine is located on the outer lip of the aperture (opening).
SIZE: To 1 5/8" (4.1 cm) high, 1" (2.5 cm) wide.
HABITAT: On semi-exposed or protected rocks and pilings from the high to mid-intertidal zone.
RANGE: Tomales Bay, Marin County, California, to Camalu, Baja California, México.
NOTES: The angular unicorn and spotted unicorn (see p.96) were once considered to be the same species. Today it is known that they are distinct. The angular unicorn lays capsules containing 40–140 eggs. Some of these eggs develop partially and then stop. They then become food for the young that have hatched from other eggs and developed normally.

MOLLUSKS AND LAMPSHELLS

SPOTTED UNICORN
Acanthina punctulata

OTHER NAMES: Spotted thorn drupe, punctata thorn drupe; formerly included with *Acanthina spirata*.
DESCRIPTION: Black spotted markings on whorls. **Shoulder of shell is well rounded**. Outer lip of aperture (opening) has a stout spine.
SIZE: To 1" (2.5 cm) high.
HABITAT: On semi-exposed rocky shores in the high intertidal zone.
RANGE: Monterey Bay, California, to Punta Santo Tomas, Baja California, México.
NOTES: The acorn barnacle (see p. 163) and checkered periwinkle (p. 85) are two main food items for the spotted unicorn, which needs a remarkable 15 to 60 hours to drill through the shell of the prey. Periwinkles have been observed to escape from this predator by crawling onto its shell—a safe place to be as long as it stays there.

CHECKERED UNICORN
Acanthina paucilirata

OTHER NAME: Checkered thorn drupe.
DESCRIPTION: Prominent **black and white checkered bands** on whorls and about 4 white spiral ribs on last whorl. Outer lip of aperture (opening) has a stout spine.
SIZE: To 1" (2.5 cm) high.
HABITAT: On rocks in the high intertidal zone.
RANGE: San Pedro, Los Angeles County, California, to Isla Cedros, Baja California, México.
NOTES: The checkered unicorn and other unicorn species (see above) obtained their common name from the presence of a unicorn-like spine at the aperture of the shell. This spine is thought to aid the snails while they dine on barnacles. This species is especially common south of San Diego, California, and is found higher on the shores than other unicorn species. Aggregations for reproduction have been observed at La Jolla, California, in March.

Leafy Hornmouth

Ceratostoma foliatum

Other Name: *Purpura foliata.*
Description: Color varies from gray to yellow or brown with **3 prominent leafy varices (frills)** extending the length of the shell. Species is noted for having pronounced **spiral ribs between the leaf-like varices**.
Size: To 3 1/2" (8.9 cm) high.
Habitat: On rocky beaches, intertidal zone to subtidal water 200' (60 m) deep.
Range: Sitka, Alaska to San Diego, California. Uncommon south of Point Conception, California.
Notes: This carnivore feeds on mussels and barnacles by drilling holes through their shells. Aggregations can often be found in late February and March when they gather to lay their yellow eggs—occasionally on the shells of their buddies.

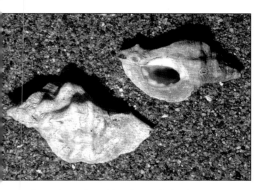

Nuttall's Hornmouth

Ceratostoma nuttalli

Other Names: Nuttall's thorn purpura; formerly *Purpura nuttalli*, *Pterorytis nuttalli*.
Description: Color varies from white to brown or white, often with spiral bands present. Three varices (flanges) extend over the shell's length. **A knobbed axial rib is present between the flanges**. A short spine protrudes near the base of the aperture, with small teeth inside.
Size: Shell to 2 1/4" (5.7 cm) high.
Habitat: On rocks from the mid- to low intertidal zone.
Range: Point Conception, Santa Barbara County, California, to Bahía Santa Maria, Baja California Sur, México.
Notes: Nuttall's hornmouth is a carnivore preferring to dine on the Pacific blue mussel (see p. 126) and the California mussel (p. 125). The scientific name for hornmouths is appropriate: *Keratos* ("spine" or "horn") and *stoma* ("mouth") for the spine located near the aperture.

MOLLUSKS AND LAMPSHELLS

GEM MUREX
Maxwellia gemma

OTHER NAMES: Maxwell's gem shell, gem rock shell.
DESCRIPTION: Whitish to grayish shell with **brown or black spiral bands**. The moderately high spired shell includes 6–7 **rounded ridges**. Aperture is round.
SIZE: To 1 3/4" (4.5 cm) high.
HABITAT: On rocks from the low intertidal zone to water 180' (55 m) deep.
RANGE: Santa Barbara, California, to Isla Asunción, Baja California Sur, México.
NOTES: The gem murex can often be found on breakwaters, among rock rubble and massive tube worm aggregations, often at the entrances to bays. The natural history of this carnivore is not well known.

FESTIVE MUREX
Pteropurpura festiva

OTHER NAMES: Festive rock shell; formerly *Shaskyus festivus, Jaton festivus*.
DESCRIPTION: Color varies from yellowish to light brown with **fine brown spiral lines**. The frilled ridges or **"wings" are rolled backwards**. A rounded ridge is positioned between each pair of frilled ridges. Aperture is oval.
SIZE: To 2 3/4" (7 cm) high.
HABITAT: On rocks, mud flats or in bays in the low intertidal zone to water 450' (135 m) deep.
RANGE: Santa Barbara, California, to Bahía Magdalena, Baja California Sur, México.
NOTES: The festive murex is a carnivore that feeds on the scaly tube snail (see p 86). It is often detected on pilings under overhanging seaweeds or on mud flats when spawning.

Frilled Dogwinkle

Nucella lamellosa

OTHER NAMES: Frilled whelk, wrinkled purple, wrinkled whelk, purple whelk, *Thais lamellosa*.

DESCRIPTION: Color ranges from white through yellow, orange, brown and purple. Heavy shell can be smooth in exposed situations and be marked by several frills in sheltered areas.

SIZE: To $3^{1}/8$" (8 cm) high.

HABITAT: On exposed and sheltered rocky beaches, low intertidal to subtidal zones.

RANGE: Aleutian Islands, Alaska, to central California.

NOTES: Large congregations of frilled dogwinkles are often found at the low tide mark in winter. At this time they breed, laying hundreds of yellow spindle-shaped eggs attached to the sides and undersides of rocks. These eggs are sometimes referred to as "sea oats." Each female can lay up to 1,000

eggs per year, but maturity is reached at 4 years. Predators of this dogwinkle include the red rock crab (see p. 182) and purple star (p. 197).

• •

Channelled Dogwinkle

Nucella canaliculata

OTHER NAMES: Channeled purple, *Thais canaliculata*.

DESCRIPTION: Color ranges from gray to yellowish. Several **grooved channels cover entire shell**.

SIZE: To $1^{1}/2$" (4 cm) high.

HABITAT: On rocky beaches, throughout intertidal zone.

RANGE: Aleutian Islands, Alaska, to Cayucos, San Luis Obispo County, California.

NOTES: The channelled dogwinkle feeds on mussels and barnacles and is often found moving among them. It drills a hole in the shell of its prey, then feeds on the delicacy inside. It has been found to require 1–2 days to complete the process of preparing (drilling) and feeding on one animal.

• •

MOLLUSKS AND LAMPSHELLS

STRIPED DOGWINKLE
Nucella emarginata

OTHER NAMES: Short-spired purple, rock-dwelling thais, rock whelk, emarginate dogwinkle; formerly *Thais emarginata*.
DESCRIPTION: Color varies from gray through yellow, brown or black, often with white bands on the ribs. **Bands on shell alternate between wide and narrow**.
SIZE: To 1" (2.5 cm) high.
HABITAT: On rocky beaches, high and mid-intertidal zones.
RANGE: Bering Sea to México.
NOTES: This is quite a variable species. Two forms of it exist: an elongated form, often found in the more northerly sites, and a rounded form, to be seen in southern locations. The striped dogwinkle is a predator that drills holes in the shells of mussels, barnacles, limpets and other snails to feed on them. The sexes of this species are separate, but both have a penis. Their yellow spindle-shaped eggs are laid in the intertidal zone.

CIRCLED ROCKSNAIL
Ocenebra circumtexta

OTHER NAMES: Circled rock snail, circled rock shell, circled dwarf triton.
DESCRIPTION: Shell white or gray with a **pair of brown spotted bands on each whorl** (a complete circling of the shell).
SIZE: To 1" (2.5 cm) high.
HABITAT: In rocky crevices from the mid- to low intertidal zone.
RANGE: Trinidad, Humboldt County, California, to Laguna Ojo de Liebre, Baja California, México.
NOTES: The circled rock snail is found in areas of heavy surf. This distinctive snail feeds on barnacles and probably other species including perhaps a variety of mollusks. White and orange color variations have also been found.

Lurid Rocksnail
Ocinebrina lurida

OTHER NAMES: Lurid rock snail, dwarf triton; formerly *Ocenebra lurida*, *Urosalpinx lurida*.
DESCRIPTION: Color varies widely including white, yellow, dark brown, red or black. Dark colored spiral bands may also be present. Up to 6 whorls are crossed with 6–10 axial ribs and fine spiral threads. Aperture has 6–7 teeth.
SIZE: Shell to 1 1/2" (4 cm) high.
HABITAT: On and under rocks on rocky exposed beaches from the low intertidal zone to water 180' (55 m) deep.
RANGE: Kenai Peninsula, Gulf of Alaska, Alaska, to Punta Santo Tomás, Baja California, México. Common north of Point Conception, Santa Barbara County, California.
NOTES: The lurid rocksnail is a common species. It feeds on barnacles, and is also known to attack the giant Pacific chiton (see p. 69), which reaches 13" (33 cm) in length. This dwarf attacks a goliath by rasping a pit in the chiton's fleshy mantle.

Poulson's Rock Snail
Roperia poulsoni

OTHER NAMES: Poulson's dwarf triton, Poulson's rock shell; formerly *Ocenebra poulsoni*.
DESCRIPTION: Shell is white with **many fine, brown spiral lines** and a semi-gloss finish. Whorls have 8–9 ribs and prominent knobs on body whorl. Aperture is oval.
SIZE: To 2 1/2" (6 cm) high, but normally much smaller.
HABITAT: On rocks from the low intertidal zone to just below the tide line.
RANGE: Cayucos, San Louis Obispo County, California, to Bahía Magdalena, Baja California Sur, México.
NOTES: Poulson's rock snail is a common species preferring sheltered bays but also found on the exposed coast. It is a carnivore that drills holes in and feeds on the California mussel (see p. 125), Pacific blue mussel (p. 126) and a variety of snails and barnacles. Its major predator is the red rock crab (p. 182) with its strong pincers.

ATLANTIC OYSTER DRILL
Urosalpinx cinerea

OTHER NAME: Eastern oyster drill.
DESCRIPTION: Exterior varies from gray to brown or yellowish; interior is tinged with purple. The shell is made up of 5–6 convex whorls and 8–12 axial ribs etched with numerous spiral cords. The canal is open.
SIZE: Shell normally to $3/4$" (2 cm) but occasionally to $1^{3/4}$" (4.4 cm) high.
HABITAT: In quiet bays and estuaries from the mid-intertidal zone to water 50' (15 m) deep.
RANGE: Boundary Bay, BC, to Newport Bay, Orange County, California.
NOTES: The Atlantic oyster drill was accidentally introduced to the Pacific with Atlantic oysters sometime between 1869 and 1888. During some years it is common at Tomales Bay, San Francisco Bay and Elkhorn Slough. It feeds on bivalves, especially mussels and oysters, as well as on barnacles. To feed, it drills a hole through the shell of its prey with a specialized boring organ. This organ produces a very thick, acidic secretion and an enzyme (carbonic anhydrase) that softens the shell, enabling the animal to drill a hole with its radula (raspy tongue) and suck out the meat.

CARINATE DOVESNAIL
Alia carinata

OTHER NAMES: Carinated dove snail, carinate dove shell, keeled dove shell, *Mitrella carinata, Columbella carinata, Nitidella carinata*.
DESCRIPTION: Color is variable from dark yellow to brown and occasionally mottled. Shell is smooth; aperture is elongated with extended teeth inside.
SIZE: Shell to $1/2$" (1.1 cm) high.
HABITAT: On rock from the intertidal zone to water 16' (5 m) deep.
RANGE: Forrester Island, southeast Alaska, to southern Baja California, México.
NOTES: The carinate dovesnail is thought to be a micro-carnivore and detritus feeder but little is known about its biology. It is often found on seaweeds, especially perennial kelp.

WRINKLED AMPHISSA
Amphissa columbiana

OTHER NAMES: Wrinkled dove snail, Columbian amphissa.
DESCRIPTION: Color varies from pink to mauve or yellow, often mottled with brown. Shell is made of several body whorls with 20–24 longitudinal ribs on the second-last whorl. The **ribs run almost parallel to the central axis of the large shell**.
SIZE: To 1 1/8" (3 cm) high.
HABITAT: On rocky beaches and mud, intertidal zone to water 96' (29 m) deep.
RANGE: Chiachi Island, Alaska, to San Pedro, Los Angeles County, California.
NOTES: This species gets its name from the ribs on the exterior of its shell. Its siphon or feeding tube is often seen extended while it searches for its next meal of carrion. This scavenger has also been found to be attracted to the food remains left by giant Pacific octopus (*Octopus dofleinki*) at its den.

VARIEGATED AMPHISSA
Amphissa versicolor

OTHER NAMES: Variegate amphissa, variegated dove shell, Joseph's coat amphissa.
DESCRIPTION: Color varies widely, often with reddish brown or white markings. Shell is made of several body whorls with approximately 15 longitudinal ribs on the second-last whorl. The **ribs run at a noticeable angle to the central shell axis**.
SIZE: To 3/4" (1.9 cm) high.
HABITAT: Often under rocks and in tidepools, from the low intertidal zone to 150' (46 m).
RANGE: Vancouver Island, BC, to northern Baja California, México.
NOTES: The variegated amphissa is commonly found in California. As its scientific name *versicolor* indicates, it occurs in many different colors, including white, gray, yellow, red and brown. Empty shells are often found washed up on the shore after a storm. Like its close relative the wrinkled amphissa (see above), this species is a scavenger and is sometimes observed with its long siphon extended as it moves about.

MOLLUSKS AND LAMPSHELLS

KELLET'S WHELK
Kelletia kelletii

OTHER NAMES: Kellett's whelk (misspelled), *K. kelleti*.
DESCRIPTION: Shell is heavy with a knobbed body whorl, normally beige, and sometimes covered with a variety of algae, bryozoans, etc.
SIZE: To $6^{1/2}$" (16 cm) high.
HABITAT: On rocks from the low intertidal zone to water 230' (70 m) deep.
RANGE: Monterey Bay, California, to Isla Asunción, Baja California Sur, México.
NOTES: Kellet's whelk has been observed feeding on dead or injured animals. It sometimes falls prey to the giant spined star (see p. 196). When it is only 7 or 8 years old, this giant reaches 3" (7.5 cm). Its shells have been found in middens of aboriginal people, who may have used them to make bird calls or trumpets. The species is also found in Japan.

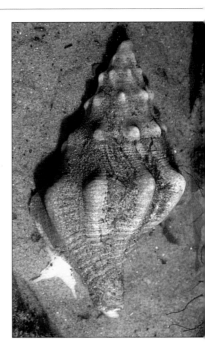

GIANT WESTERN NASSA
Nassarius fossatus

OTHER NAMES: Channeled dog whelk, channeled basket snail.
DESCRIPTION: Shell color ranges from gray to orange-brown; interior edge is a distinctive orange. The 7 whorls have many fine longitudinal ridges.
SIZE: To 2" (5 cm).
HABITAT: In sheltered areas under rocks or buried in sand and mud, intertidal zone to water 60' (18 m) deep.
RANGE: Queen Charlotte Islands, BC, to Laguna San Ignacio, Baja California, México.
NOTES: This carnivore drills holes in clam and snail shells to feed on the inner tissues, using its long proboscis (snout), which is a feeding organ. Its distinctive shell can sometimes be found on the beach.

Western Lean Nassa
Nassarius mendicus

Other Names: Lean nassa, *N. cooperi*.
Description: Yellowish or brown in color with white spiral banding. Shell is slender with strong axial ribs crossed by flat spiral cords.
Size: Shell to 7/8" (2.2 cm) high.
Habitat: On sand or mud from the low intertidal zone to water 200' (61 m) deep.
Range: Kodiak Island, Gulf of Alaska, to Isla Asunción, Baja California Sur, México.
Notes: Unlike other related species, the western lean nassa crawls about on sand rather than under it. This snail is found on both the open coast and in bays. The number of ribs on each whorl varies considerably, from just a few on shells occurring in southern California to 12 in northern populations.

Western Fat Dog Nassa
Nassarius perpinguis

Other Names: Western fat nassa, fat nassa.
Description: Shell is yellowish with 2 or 3 narrow bands of orange-brown. Spiral threads with beaded ridges cover the exterior of the shell.
Size: Shell to 1" (2.5 cm) high.
Habitat: On sandy bottoms from the low intertidal zone to water 300' (90 m) deep.
Range: Point Reyes, Marin County, California, to Isla Cedros, Baja California, México.
Notes: The western fat dog nassa (*pinguis* means "fat") is more common subtidally than intertidally in lagoons, bays and offshore.

MOLLUSKS AND LAMPSHELLS

EASTERN MUD SNAIL
Ilyanassa obsoleta

OTHER NAMES: Eastern mud whelk, eastern mud nassa, black dog whelk, mud dog whelk, worn out dog whelk; formerly *I. obsoletus, Nassarius obsoletus*.
DESCRIPTION: Shell is dark brown to black, sometimes with a white spiral band. It is spindle-shaped with 6 whorls, a blunt end (often eroded) and several spiral threads. A greenish color is sometimes present due to microscopic algae.
SIZE: To $1^{1/8}$" (3 cm) high.
HABITAT: On mud flats from the low intertidal zone to just below it.
RANGE: Vancouver Island, BC, to Monterey Bay, California.
NOTES: This species was introduced from the Atlantic coast before 1911. It feeds on both organic material and carrion, and buries itself in the mud when the tide goes out. Under the right conditions it can be found in accumulations numbering into the thousands. It is thought to have a life span of 5 years. Dark individuals are probably the most unattractive snails found along the Pacific coast.

PAINTED SPINDLE
Fusinus luteopictus

OTHER NAMES: Painted spindle shell; formerly *Aptyxis luteopicta*.
DESCRIPTION: Shell is dark brown with a yellowish white spiral band in the middle of each whorl. The spindle-shaped shell holds several convex whorls, each of which includes 9–10 strong axial ribs and 2 strong spiral chords that cross them, creating elongated beads.
SIZE: Shell to 1" (2.5 cm) high.
HABITAT: On and under rocks from the low intertidal zone to water 130' (40 m) deep.
RANGE: Monterey Bay, California, to Isla San Gerónimo, Baja California, México.
NOTES: The painted spindle lays its eggs in capsules, which it cements to rocks. Tiny spiral shells of a serpulid marine worm are often found attached to the shell of the painted spindle.

PURPLE OLIVE
Olivella biplicata

OTHER NAME: Purple dwarf olive.
DESCRIPTION: Color varies from **near-white to lavender** and dark purple. Shell is smooth and shiny, indicating that a large, fleshy foot is present.
SIZE: To 1 1/4" (3 cm) high.
HABITAT: On exposed sandy beaches, low intertidal zone to water 150' (46 m) deep.
RANGE: Sitka, Alaska, to Bahía Magdalena, Baja California Sur, México.

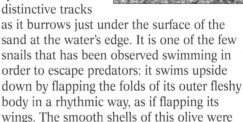

NOTES: This species leaves distinctive tracks as it burrows just under the surface of the sand at the water's edge. It is one of the few snails that has been observed swimming in order to escape predators: it swims upside down by flapping the folds of its outer fleshy body in a rhythmic way, as if flapping its wings. The smooth shells of this olive were admired and used as money and ornaments by aboriginal peoples, in whose graves strings of shells have been found.

BAETIC OLIVE
Olivella baetica

OTHER NAME: Little olive.
DESCRIPTION: **Creamy with reddish brown markings**, often with purple or brown bands around each whorl. Shell is an elongated oval.
SIZE: To 3/4" (1.9 cm) high.
HABITAT: On exposed and sheltered sandy beaches, low intertidal zone to water 204' (62 m) deep.
RANGE: Kodiak Island, Alaska, to Cabo San Lucas, Baja California Sur, México.
NOTES: A similar species, the purple olive (above), is a wider and larger snail often found on the same beaches. Like all *Olivella*, the baetic olive has no eyes: it burrows in the sand most of its life so it does not need to see.

MOLLUSKS AND LAMPSHELLS

ZIGZAG OLIVE
Olivella pycna

DESCRIPTION: Shell color varies from fawn to brown with prominent **zigzag markings** and a dull luster. The shell is stout and chunky.
SIZE: To $5/8"$ (1.5 cm) long.
HABITAT: On sandy shores from the low intertidal zone to water 90' (27 m) deep.
RANGE: Vancouver Island, BC, to Monterey Bay, California.
NOTES: This beautiful species is found more often in the southern portion of its range, especially at sites like Bodega Bay and Humboldt Bay, northern California. Populations of zigzag olives fluctuate greatly, from abundant in some years to absent or nearly absent in others; the cause is unknown. Hermit crabs sometimes take up residence in the empty shells of this small snail.

IDA'S MITER
Mitra idae

OTHER NAMES: Half-pitted miter, *M. catalinae, M. diegensis, M. montereyi.*
DESCRIPTION: Shell is dark brown with a black periostracum (covering), elongate and heavy with fine axial and spiral lines. The inner margin of the aperture has 3 folds. The body is striking white.
SIZE: Shell to $3 1/8"$ (8 cm) high.
HABITAT: Among rocks from the low intertidal zone to water 60' (18 m) deep.
RANGE: Crescent City, Del Norte County, California, to Isla Cedros, Baja California, México.
NOTES: Ida's miter deposits its eggs in elongated, flattened capsules, yellow in color or transparent. Each capsule can contain 100 to nearly 1000 eggs. During the spawning period, up to 264 capsules may be deposited on rocks. This carnivore hides under seaweed during the day and hunts by night.

Banded California Marginella
Volvarina taeniolata

Other Names: California marginella, California margin shell; formerly *Marginella californica, Hyalina californica*.
Description: Color varies from ivory to yellow-orange with light bands. The shell is shiny; aperture is elongated and has 4 folds on its inner margin.
Size: Shell to 1/2" (1 cm) high.
Habitat: On undersides of rocks from the low intertidal zone to water 150' (46 m) deep.
Range: Point Conception, Santa Barbara County, California, to Salinas, Guayas Province, Ecuador.
Notes: This small, common species is often called "wheat shells" because of its size and shape. Little is known about its natural history.

California Cone
Conus californicus

Other Names: California cone shell, *Conus ravus*.
Description: Grayish brown shell with a chestnut-colored periostracum (covering). Shell is conical with a narrow opening.
Size: To 1 5/8" (4.1 cm) high.
Habitat: On rocky and sandy bottoms from the low intertidal zone to water 100' (30 m) deep.
Range: Farallon Islands, west of San Francisco, California, to Bahía Magdalena, Baja California Sur, México.
Notes: The California cone is a predator of many invertebrates, including the purple olive (see p. 107), California jackknife clam (p. 141) and several species of segmented worms. Native people used the empty shells of the California cone as money and as ceremonial items. Cones are a well-known group of snails, partly because the poison of some large Indo-Pacific species is known to be lethal to humans.

MOLLUSKS AND LAMPSHELLS

GRAY SNAKESKIN-SNAIL
Ophiodermella inermis

OTHER NAMES: Formerly *O. ophioderma, O. incisa*.
DESCRIPTION: Shell is gray with fine black lines, and slender. There is an anal notch at the widest portion of the aperture.
SIZE: Shell to 1 1/2" (4 cm) high.
HABITAT: From the low intertidal zone to water 230' (70 m) deep.
RANGE: Skidegate, Queen Charlotte Islands, BC, to Rancho Inocentes, Baja California Sur, México.
NOTES: This snail feeds on polychaete worms. Because it cannot detect its prey chemically from any distance, it moves against the current until it reaches its quarry. The red rock crab (see p. 182) is a major predator in the intertidal zone. Gray snakeskin-snails reproduce from October to July. The egg capsules are usually laid in February, and each one holds several hundred eggs. The beautiful shells of this snail are sometimes found washed up on the beach after a storm.

CALIFORNIA BUBBLE
Bulla gouldiana

OTHER NAMES: Cloudy bubble snail, Gould's bubble, Gould's bubble shell.
DESCRIPTION: Body is yellowish brown or orange with white speckles; shell is reddish or grayish brown mottled with dark, angular shapes and white borders.
SIZE: Shell to 2 1/4" (5.5 cm) long.
HABITAT: On mud and sand in the low intertidal zone.
RANGE: Morro Bay, San Luis Obispo County, California, to Ecuador.
NOTES: This species feeds on detritus in the mud and sand of bays and the sheltered outer coast. It is believed to live only 1 year.

WHITE BUBBLE
Haminoea vesicula

OTHER NAMES: White bubble shell, white bubble snail, blister glassy-bubble.
DESCRIPTION: Light brown; shaped somewhat like a slug with a thin shell covering a portion of the body. **Aperture size is less than half the diameter of the shell.**
SIZE: To 3/4" (2 cm) high.
HABITAT: On eel-grass and seaweed in muddy bays or sandy areas, intertidal zone. Also known to burrow just below surface of sand or mud.
RANGE: Ketchikan, Alaska, to Bahía Magdalena, Baja California Sur, México.
NOTES: The internal shell of this species is almost completely covered by its body. Its deep yellow egg ribbons can sometimes be seen in July, on sandy-bottom shores. If you are lucky enough to find a particularly light specimen of this species, it is possible to observe the pulsing of its heart through the shell.

GREEN BUBBLE
Haminoea virescens

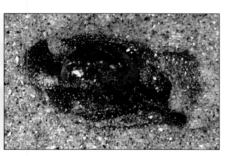

OTHER NAMES: Green paper bubble, Sowerby's paper bubble, *H. cymbiformis*, *H. strongi*.
DESCRIPTION: Light green; shaped somewhat like a slug with a thin shell. **Aperture size is more than half the diameter of the shell.**
SIZE: To 3/4" (2 cm) long.
HABITAT: This species is found on the open coast and occasionally in bays in the low intertidal zone.
RANGE: Southern Alaska to Bahía de Panama, Panama.
NOTES: The green bubble shell is occasionally seen in abundance on open rocky shores. It has been found at White's Point, Point Fermin and Crystal Cove, southern California.

MOLLUSKS AND LAMPSHELLS

NUDIBRANCHS AND ALLIES
Subclass Opisthobranchia and Gymnomorpha

The nudibranchs, or "sea slugs," are favorites of divers, beachcombers and snorkelers because many of them display such spectacular colors and patterns. Others have coloring that allows them to match their environments very closely. The nudibranch has a shell early in its life, but the shell is soon lost. Predators are few and far between. Many nudibranchs have chemical defenses or discharge stinging cells called nematocysts. All nudibranchs have a pair of intricate projections near the head, called rhinophores, which help them detect chemicals in the water. Some of these chemicals can help lead the nudibranch to food sources.

Nudibranchs
Subclass Opisthobranchia

CALIFORNIA SEA HARE
Aplysia californica

OTHER NAMES: California brown sea hare, sea hare, *Tethys californica*.
DESCRIPTION: Color varies from reddish to brown or greenish with a wide variety of mottling. 2 elongated flaps are found on either side of the body. 2 pairs of antennae are present. A small shell is found internally.
SIZE: To 16" (40 cm) long and 8" (20 cm) wide.
HABITAT: Sheltered sites from the low intertidal zone to water 60' (18 m) deep.
RANGE: Yaquina Bay, Oregon, to Baja California, México.
NOTES: Sea hares get their name from one pair of antennae, which resemble a hare's ears. The California sea hare is a herbivore, grazing primarily on red algae. It can discharge a dark purple ink if irritated. Each adult is a hermaphrodite, having both male and female reproductive organs, but each must locate another individual to mate. One 5 lb 12 oz (2.6 kg) individual was observed to lay some 478 million eggs over a 17-week period. The spaghetti-like eggs are yellowish green in color.

California Black Sea Hare

Aplysia vaccaria

OTHER NAME: Black sea hare.
DESCRIPTION: This species is very similar to the California sea hare (see p.112). Color is normally uniform black to dark brown, sometimes with fine white or gray speckles.
SIZE: To 40" (1 m) long, but intertidal specimens much smaller.
HABITAT: On rocky coasts and kelp beds from the low intertidal zone to subtidal waters.
RANGE: Monterey Bay, California, to Gulf of California.
NOTES: The California black sea hare is the world's largest gastropod, weighing in at about 31 lbs (14 kg). Unlike the California sea hare, it does not produce a purple ink-like secretion when disturbed. The main food of this quiet herbivore is the feather boa kelp (see p. 231). Its pinkish white eggs are deposited in tangled strings.

Egg cluster of California black sea hare.

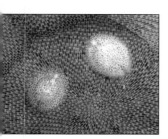

Cryptic Nudibranch

Doridella steinbergae

OTHER NAME: Steinberg's dorid.
DESCRIPTION: Color is a mixture of brown and white in an irregular grid pattern. Body is tiny and flat, closely resembling the bryozoan on which it feeds.
SIZE: To 5/8" (1.7 cm) long.
HABITAT: Usually found feeding on kelp encrusting bryozoan (see p. 190), low intertidal zone.
RANGE: Prince William Sound, Alaska, to Baja California, México.
NOTES: This nudibranch is easy to miss unless you are looking for it. Its only food source, the kelp encrusting bryozoan, lives on bull kelp (see p. 233) and a few other seaweeds. Check the bryozoan closely, as the tiny cryptic nudibranch spends most of its life there.

HOPKIN'S ROSE
Hopkinsia rosacea

OTHER NAME: Rose nudibranch.
DESCRIPTION: Deep rosy pink dorid with elongated papillae (finger-like projections) covering its upper body.
SIZE: To 3/4" (2 cm) long.
HABITAT: On rocky shores from the low intertidal zone to water 20' (6 m) deep.
RANGE: Coos Bay, Oregon, to Isla San Martin, Baja California, México.
NOTES: Hopkin's rose feeds on the derby hat bryozoan (see p. 191) by drilling a hole in its outer crust and sucking out the soft inner tissues. Its eggs are laid in a spiraling, pink to rose-colored ribbon laid in a counter-clockwise pattern. This seasonally common nudibranch secretes substances that apparently repel predators.

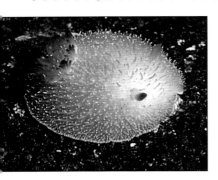

RUFUS TIPPED NUDIBRANCH
Acanthodoris nanaimoensis

OTHER NAMES: Nanaimo dorid, *Acanthodoris columbina*.
DESCRIPTION: White or gray body covered with yellow-tipped projections. Red tips grace the antennae-like rhinopores and edges of the gills.
SIZE: To 1 1/8" (3 cm) long.
HABITAT: Low intertidal zone to subtidal waters.
RANGE: Halibut Point, Baranof Island, Alaska, to Shell Beach, San Luis Obispo County, California.
NOTES: There are two distinct color phases for this species. In the light phase the body is typically white overall, while the darker form is mottled with dark gray. In both phases, distinctive yellow-tipped projections cover the entire body and red is present on the gills and the tips of the rhinophores. This nudibranch feeds on bryozoans.

Barnacle-Eating Dorid

Onchidoris bilamellata

OTHER NAMES: Rough-mantled sea slug; formerly *Onchidoris fusca*.
DESCRIPTION: Body is a mixture of browns and cream colors, and is entirely covered by blunt-tipped projections. The simple gills form a broad horseshoe.
SIZE: To 3/4" (2 cm) long.
HABITAT: On rocks on or near mud bottoms, low intertidal zone.
RANGE: Hagemeister Island, Alaska, to Cape Colnett, Baja California, México.
NOTES: Barnacles are the only food of the adults of this species, which suck out the barnacle's body contents with a special pumping mechanism. (Juveniles feed on encrusting bryozoans.) This dorid also occurs on both sides of the Atlantic coast.

Orange-Spotted Nudibranch

Triopha catalinae

OTHER NAMES: Sea clown triopha, *Triopha carpenteri*.
DESCRIPTION: White, elongated body with bright orange or red tips on all projections. The animal has rhinophores (antennae-like projections) and a circlet of branched gills.
SIZE: Normally to 2 3/4" (7 cm) long, but can grow to 6" (15 cm) long.
HABITAT: In tidepools and kelp beds, mid-intertidal zone to water 115' (35 m) deep.
RANGE: Amchitka Island, Alaska, to El Tomutal, Baja California, México.
NOTES: This nudibranch and other species have been observed crawling upside down on the surface of tidepools. The individual can do this by secreting a trail of slime and moving along it. If it falls, it merely crawls back to resume its course. This species feeds on bryozoans.

MOLLUSKS AND LAMPSHELLS

SPOTTED TRIOPHA
Triopha maculata

OTHER NAME: Includes *T. grandis*
DESCRIPTION: Body color varies widely from yellow to orange, red or brown. Small white and blue oval spots cover the elongated body and orange-red processes are found on the oral veil and sides.
SIZE: To 2" (5 cm) long.
HABITAT: In tidepools from the low intertidal zone to water 110' (33 m) deep.
RANGE: Bamfield, Vancouver Island, BC, to Punta Cono, Baja California, México.
NOTES: The spotted triopha feeds on several bryozoans. Its egg ribbons can be seen on kelp in subtidal waters. Smaller sized juveniles are normally found intertidally.

COCKERELL'S DORID
Laila cockerelli

OTHER NAME: Cockerell's nudibranch.
DESCRIPTION: Body color varies from white to yellowish with numerous long, orange-tipped papillae (projections). A small white gill circlet is also present.
SIZE: To 1" (2.6 cm) long.
HABITAT: In tidepools from the low intertidal zone to water 115' (35 m) deep.
RANGE: Vancouver Island, BC, to Cabo San Lucas, Baja California, México.
NOTES: The colorful Cockerell's dorid dines on bryozoans. It lays pale pink eggs in long spirals that contain about 6,500 eggs.

Yellow-Edged Nudibranch

Cadlina luteomarginata

OTHER NAMES: Yellow-edged cadlina, *Cadlina marginata*.
DESCRIPTION: White body edged with bright lemon-yellow trim. A ring of 6 frilly white gills tipped with yellow are located at the rear.
SIZE: To 3" (8 cm) long.
HABITAT: Under ledges and rocks in tidepools, low tide to water 65' (20 m) deep.
RANGE: Alaska to Pt. Eugenia, México.
NOTES: The body of this beautiful nudibranch feels gritty or sandpaper-like because of the many sharp projections on its dorsal surface. The yellow-edged nudibranch feeds on a variety of sponge species.

Red Nudibranch

Rostanga pulchra

OTHER NAMES: Crimson doris, red sponge nudibranch.
DESCRIPTION: Bright red or tan, similar in color to the red sponge on which it feeds. Small oval body, rounded in front and pointed toward the rear.
SIZE: To 1 1/8" (3 cm) long, but intertidal specimens tend to be much smaller.
HABITAT: On rocky shores with overhanging ledges or large boulders, intertidal zone to water 30' (9 m) deep.
RANGE: Alaska to Gulf of California, México.
NOTES: This common nudibranch is an excellent example of camouflage. It incorporates into its own body the pigment of the red sponges on which it feeds, which helps it to blend in to the red sponge substrate.

Its ring-like egg masses are also vivid red and laid on sponges in spring or early summer, so that they too blend in with the substrate. The species' whole life revolves around the red sponges on which its feeds. Studies have shown that some individuals move about a great deal, while others are much more sedentary. One individual was observed to have stayed on a sponge for 37 days.

MONTEREY DORID
Archidoris montereyensis

OTHER NAME: Monterey sea lemon.
DESCRIPTION: Flat body, shaped somewhat like a slice of lemon with **small black spots covering some of its small rounded body projections**. Two horn-like rhinophores at front; cluster of yellowish brown gills at rear.
SIZE: To $2^{3/4}$" (7 cm) long on average, but can grow to $4^{3/4}$" (12 cm) long.
HABITAT: On rocky shorelines, intertidal zones to water 160' (49 m) deep.
RANGE: Kachemak Bay, Alaska, to San Diego, California.
NOTES: The Monterey dorid lays its eggs on rock, year round. Each spiral-shaped cluster may contain up to 2 million eggs. The bread crumb sponge (see p. 18) is the primary food and this dorid can often be found there. A similar species, the sea lemon (below), can be distinguished by the black spots between the small projection tips on its back or dorsal side.

SEA LEMON
Anisodoris nobilis

OTHER NAMES: Noble Pacific doris, speckled sea lemon.
DESCRIPTION: Flat, bright yellow to orange body covered dorsally with short rounded projections, sometimes with **black spots between the projections**. A pair of pointed rhinophores (antennae-like projections) at head region, a pair of frilly white gills at rear.
SIZE: Usually to 4" (10 cm) long, but occasionally to 10" (26 cm).
HABITAT: Low intertidal zone to water 750' (230 m) deep.
RANGE: Kodiak Island, Alaska to Baja California, México.
NOTES: This common nudibranch feeds on several types of encrusting sponges. Many nudibranchs have a distinctive smell; the sea lemon has a somewhat fruity odor, which may help repel predators. Like other dorids, this one has a flattened body and lacks large cerata (projections) along its body. The gills are usually seen as a rosette on the animal's back.

White Spotted Sea Goddess

Doriopsilla albopunctata

OTHER NAMES: White spotted porostome, yellow porostome, white-speckled nudibranch, salted doris, salted yellow doris; includes *Dendrodoris fulva*.

DESCRIPTION: Body color is yellow or yellow-orange to reddish brown overall with **white gills. White spots are found only on the tips of the tubercles** (small projections).
SIZE: To 2 1/2" (6 cm) long.
HABITAT: On rocky shores from the low intertidal zone to 150' (45 m).
RANGE: Mendocino, California, to Point Eugenia, México; northern Gulf of California.
NOTES: The white spotted sea goddess is a well-named treasure for beachcombers and divers alike. Its color is truly stunning! Darker colored individuals are more common in the southern portion of its range.

Yellow-gilled Sea Goddess

Doriopsilla gemela

OTHER NAMES: Yellow-gilled porostome.
DESCRIPTION: Body is **yellow to yellow-orange or orange-brown**, including rhinophores (antennae-like projections) and **gills**. White spots are found both on and off the tubercles (small projections).
SIZE: To 1 1/2" (4 cm) long.
HABITAT: In tidepools in the low intertidal zone and subtidal depths of 16' (5 m).
RANGE: Elkhorn Slough, Monterey County, California, to Bahía de Tortuga, Baja California, México; northern Gulf of California.
NOTES: The yellow-gilled sea goddess is a near twin to the white spotted sea goddess (above), as the scientific name *gemela* ("twin") suggests. This is one of few nudibranchs that have no radula (raspy tongue). Instead, the porostomes feed on sponges by releasing a fluid and sucking up the partially digested food with a special tube located at the mouth.

MOLLUSKS AND LAMPSHELLS

RINGED NUDIBRANCH
Diaulula sandiegensis

OTHER NAMES: Spotted nudibranch, ringed doris, brown-spotted nudibranch, leopard nudibranch, San Diego dorid, ring-spotted nudibranch.
DESCRIPTION: Light brown to gray with various sizes of brown rings or spots distributed randomly over the body. A pair of white antennae in head region, a pair of frilly gills at rear.

SIZE: Normally to 2 3/4" (7 cm) long, but has been known to grow to 4" (10 cm).
HABITAT: Under rocks, ledges and kelp, low intertidal zone to subtidal waters 110' (34 m) deep.
RANGE: Alaska to Cabos San Lucas, Baja California, México.
NOTES: The ringed nudibranch feeds on intertidal and subtidal sponges. An overhanging rock ledge is a typical spot for this animal to lay its white spiral of eggs—up to 16 million of them. Northern specimens are generally darker and have more rings or spots on their body than those from the south.

WHITE DENDRONOTID
Dendronotus albus

OTHER NAMES: White dendronotus.
DESCRIPTION: Color white overall, usually with orange tips on gill tufts and front appendages. A bright white line is found on the dorsal surface of the tail.
SIZE: To 1 1/2" (3.5 cm) long.
HABITAT: On kelp or in tidepools from the low intertidal zone to water 100' (30 m) deep.
RANGE: Kenai Peninsula, Alaska, to Los Coronados Islands, Baja California, México.
NOTES: This small nudibranch prefers areas with a swift current, where it feeds on hydroids. The graceful tufts (gills) of this species may range in color from all white to half-covered with bright orange. In the southern part of its range, it is uncommon during the summer months.

Bushy-Backed Nudibranch

Dendronotus frondosus

OTHER NAMES: Bushy-backed sea slug, *D. arborescens, D. venustus.*
DESCRIPTION: Color varies greatly from white to grayish brown or rusty red. Two rows of bushy projections are positioned on its back in a "comb."
SIZE: To 2 1/2" (6 cm) long.
HABITAT: From the low intertidal zone to water 360' (110 m) deep.
RANGE: Alaska to California.
NOTES: The bushy-backed nudibranch is often found in bays and on floats, especially during the summer months. It feeds primarily on hydroids. This common nudibranch is cosmopolitan throughout the Northern Hemisphere. *Dendronotus* means "tree-back," referring to the bushy appearance of its back.

Frosted Nudibranch

Dirona albolineata

OTHER NAMES: Alabaster nudibranch, white-streaked dirona, white-lined dirona, chalk-lined dirona.
DESCRIPTION: Opaque, with leaf-like projections that are edged with a fine, brilliant white line.
SIZE: Normally to 1 1/2" (4 cm), but can grow to 7" (18 cm) long.
HABITAT: On rocky shores, low intertidal zone to water 100' (30 m) deep.
RANGE: Kachemak Bay, Alaska, to San Diego, California.
NOTES: This nudibranch feeds on small snails by cracking the shells with its powerful jaws. Other foods include sea anemones, sea squirts and bryozoans. The frosted nudibranch is often found during very low tides. It occurs in 3 color phases: white, purple-tinged and orange-tinged.

PAINTED NUDIBRANCH
Dirona picta

OTHER NAME: Spotted dirona.
DESCRIPTION: Color of body varies widely from pinkish orange to light brown or grayish green. In addition a wide range of colored spots are found

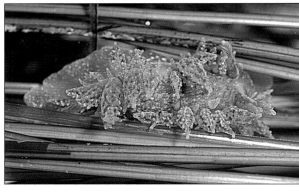

scattered over the body. Cerata (finger-like projections) are large, numerous and flattened. A red spot is often found on the lower surface of the cerata.
SIZE: To 1 1/3" (3 cm) long.
HABITAT: On rocky shores from the low intertidal zone to water 30' (9 m) deep.
RANGE: Charleston, Oregon, to Guardian Angel Island, Baja California, México.
NOTES: The painted nudibranch is aptly named with its wide-ranging colors. It is sometimes found in large numbers. The species feeds on bryozoans and has been observed moving upside-down along the surface of a tidepool.

SPANISH SHAWL
Flabellina iodinea

OTHER NAMES: Purple fan nudibranch, elegant eolid; formerly *Flabellinopsis iodinea, Coryphella sabulicola.*
DESCRIPTION: Body is deep purple overall and cerata (finger-like projections) are purple with vivid orange tips. The rhinophores (antennae) are deep maroon.
SIZE: To 1 1/2" (4 cm) long.
HABITAT: On pilings or kelp from the mid-intertidal zone to water 120' (36 m) deep.
RANGE: Vancouver Island, BC, to Bahía San Quintin, Baja California, México; northern Gulf of California.
NOTES: Spanish shawl is a strikingly colorful species, whose bright colors warn predators of its bad taste. This nudibranch can swim by making a series of quick lateral, U-shaped bends to its body, first to one side, then to the other. Spanish shawl feeds on the stick hydroid *Eudendrium ramosum.*

Three Lined Aeolid
Flabellina trilineata

OTHER NAMES: Three-lined nudibranch; formerly *Coryphella trilineata, C. fisheri*.
DESCRIPTION: Body varies from transparent white to gray with **three bright white lines** extending **over the length of the body**. Many red or orange cerata (projections) with white tips protrude from the body.
SIZE: To 1 1/2" (3.6 cm) long.
HABITAT: On rock from the low intertidal zone to subtidal waters.
RANGE: Lisianski Inlet, Alaska, to Bahía Tortugas, Baja California, México.
NOTES: The three lined aeolid is one of several species that extract unexploded nematocysts (stinging cells) from hydroids and store them in the cerata (finger-like projections) for their own protection. This nudibranch feeds on hydroids. It lays its eggs in distinctive spiral loops, which it places on seaweeds.

Shag-Rug Nudibranch
Aeolidia papillosa

Shag-rug nudibranch and eggs in tidepool.

OTHER NAMES: Shaggy mouse nudibranch, maned nudibranch, papillose aeolid.
DESCRIPTION: Whitish or pinkish body with gray or brown spots. Many colorless or gray-brown cerata (projections), typically with a bald spot down the middle.
SIZE: To 2 3/4" (7 cm) long.
HABITAT: On protected rocky shores, sandy beaches and in eel-grass beds, intertidal zone to water 2,200' (671 m) deep.
RANGE: Alaska to southern California.
NOTES: Sea anemones, including the plumose anemone (see p. 34) and the aggregating anemone (p. 29), provide food for this appropriately named nudibranch. It feeds at least once a day, consuming up to 100% of its body weight. In order for the shag-rug nudibranch to feed on anemones, it must first become immune to the stinging properties of the nematocysts (stinging cells) on the anemone's tentacles. Apparently the nudibranch first touches the anemone, then retreats, which activates the production of a coating resistant to the stinging effects. The nudibranch can then feed on the anemone without becoming a meal itself.

MOLLUSKS AND LAMPSHELLS

OPALESCENT NUDIBRANCH
Hermissenda crassicornis

OTHER NAMES: Hermissenda nudibranch, thick-horned nudibranch, long-horned nudibranch, *Phidiana crassicornis*.
DESCRIPTION: Colorful yellow-green body, orange areas on the back and a clear blue line on the sides. White tips of the cerata (projections) contain nematocysts (stinging cells) ingested from hydroids while feeding.
SIZE: Normally to 1 1/2" (4 cm) long, but can grow to 3" (8 cm) long.
HABITAT: On mud flats, eel-grass beds, docks and rocky intertidal areas.
RANGE: Alaska to Punta Eugenia, México.
NOTES: This nudibranch feeds on hydroids, sea squirts, ascidians, other mollusks, various types of eggs and pieces of fish. It is an aeolid, a type of nudibranch with many cerata (projections) along its body. The appendages in the center are responsible for digestion, while the surface ones are used for respiration. As with other aeolids, this one extracts the stinging cells from hydroids and anemones while feeding, then stores the cells in the cerata and uses them to repel predators.

Nudibranch Allies
Subclass Gymnomorpha

RETICULATE BUTTON SNAIL
Trimusculus reticulatus

OTHER NAMES: Button shell, reticulate gadinia; formerly *Gadinia reticulata*.
DESCRIPTION: White shell with radiating ribs from low apex. Eyes are located at the base of the tentacles. Underside of the **empty shell bears a horseshoe-shaped scar.**
SIZE: Shell to 3/4" (2 cm) in diameter.
HABITAT: On rocks in the mid-intertidal zone.
RANGE: Olympic Peninsula, Washington, to Acapulco, México.
NOTES: The reticulate button snail is often overlooked because of its secretive habits. It is a locally common species that lives in colonies and hides in crevices, caves and in the abandoned burrows of rock-boring bivalves. This air-breathing snail lives above the water more than below and often remains in one place for extended periods of time—sometimes a number of months.

BIVALVES
Class Bivalvia

Bivalves are a large group of animals each of which is covered by a pair of shells (valves). There are some 10,000 species of bivalves worldwide, including mussels, oysters, clams, scallops and shipworms. Some live in fresh water, but the majority are found in salt water environments, buried in sand and mud (clams), rock (piddocks) and wood (shipworms). Others have evolved to become mobile and capable of swimming (scallops). Bivalves have come to occupy very diverse environments, but remain relatively unchanged through time.

The world's largest bivalve, from the Indo-Pacific, grows to 5' (1.5 m) long and can weigh as much as 650 lbs (295 kg). Locally, the Pacific geoduck (see p. 153) can weigh up to 20 lbs (9 kg), while other species are as small as $1/16$" (2 mm) wide. Clams feed, breathe and expel wastes through siphons, special tubes that extend from the clam to the surface. The siphons range from very short, as in the Nuttall's cockle (p. 137), to 3' (1 m) long, as in the Pacific geoduck. Many species of bivalves have been, and continue to be, utilized for food.

CALIFORNIA MUSSEL
Mytilus californianus

OTHER NAMES: Surf mussel, sea mussel, ribbed mussel.
DESCRIPTION: Thin blue-black periostracum (covering) over shells. Exterior often has a streak of brown. A series of **rounded ridges extend the length of each shell**.
SIZE: To 10" (25 cm) long.
HABITAT: On exposed rocky shores, mid-intertidal zone to depths of 330' (100 m).
RANGE: Alaska to Punta Rompiente, Baja California Sur, México.
NOTES: This species has a high reproduction capability: 100,000 eggs can be produced annually by one female. The California mussel can grow $3^{1}/_{2}$" (9 cm) in one year's time. It also has many predators, including sea stars, predaceous snails, crabs, gulls, sea otters and humans. Aboriginal people of the west coast used this species extensively, its flesh for food and its shells for implements. A similar species, the Pacific blue mussel (below), is smaller and lacks the rounded ridges.

MOLLUSKS AND LAMPSHELLS

PACIFIC BLUE MUSSEL
Mytilus trossulus

OTHER NAMES: Edible mussel, blue mussel, bay mussel, foolish mussel; formerly (and incorrectly) *Mytilus edulis*.
DESCRIPTION: Color varies from blue or brown to black. Distinctive **smooth, wedge-shaped shells, narrow (anterior) end is curved**.
SIZE: To $4^{1}/_{2}$" (11 cm) long, usually much smaller.
HABITAT: In sheltered areas, mid-intertidal to subtidal water 132' (40 m) deep.
RANGE: Alaska to central California and possibly southern California.
NOTES: This mussel produces a sticky substance that hardens into thread-like fibers, the byssus, with which the mussel attaches to substrates such as rocks. If you try to remove a mussel from its bed, you will discover how strong these fibers can be. Mussels feed on plankton by pumping up to 3 quarts (3 L) of seawater per hour through their gills. This flow is generated by the rhythmic movement of cilia (tiny hairs). When the tide goes out, mussels close their shells tight to stop from drying out. This species has been studied extensively, and its former scientific name *Mytilus edulis* has been recently revised due to results of electrophoretic studies (a specialized technique of measuring suspended particles moving in fluid toward an electrode).

MEDITERRANEAN MUSSEL
Mytilus galloprovincialis

OTHER NAMES: Formerly *Mytilus edulis diegensis* and many others worldwide.
DESCRIPTION: Shell is dark blue with a light-colored hinge. Triangular **shell is broad; the narrow (anterior) end is pointed** rather than curved. Anterior muscle scar is tiny and posterior scar is broad.
SIZE: To 6" (15 cm) long.
HABITAT: On rocks from the intertidal zone to water 15' (5 m) deep.
RANGE: Primarily from central California to Manzanillo, Colima, México, but full northern range extends to southern BC.
NOTES: This mussel was introduced after 1900, from the Mediterranean as well as Britian, France and Ireland. It is currently being cultured commercially in Puget Sound, Washington. This species replaces the Pacific blue mussel (see above) from southern California to Baja California, but the two mussels cannot be distinguished by their shells alone, and they hybridize readily.

RIBBED MUSSEL
Geukensia demissa

OTHER NAMES: Ribbed horse mussel, ribbed horsemussel, Atlantic ribbed mussel, northern ribbed mussel; formerly *Mytilus demissus*, *Modiola plicatula*, *Modiola semicostata*.

DESCRIPTION: The elongated shells are coated with a thick, dark brown to black periostracum (covering), which is often worn through in spots to reveal grayish white to silvery shell exterior. **Prominent radiating ribs** can be seen on the surface of the shell.

SIZE: To 5" (12.7 cm) long.

HABITAT: Partially buried in mud or attached to pilings in salt marshes and bays from the high to mid-intertidal zone.

RANGE: San Francisco Bay, California, to upper Newport Bay, Orange County, California.

NOTES: The ribbed mussel was introduced with the Atlantic oyster (see p. 130) to San Francisco Bay before 1890. This remarkable species has a high range of tolerances including dehydration, temperatures to 133°F (56°C) and salinities twice that of normal seawater. The clapper rail (*Rallus longirostris*) is known to feed on the ribbed mussel, and occasionally the shells close on the bird's toes, causing it to drown when the tide comes in.

BRANCH-RIBBED MUSSEL
Septifer bifurcatus

OTHER NAMES: Bifurcate mussel, branched ribbed mussel, platform mussel.

DESCRIPTION: Shell exterior whitish yellow with a black periostracum (covering) and **prominent radiating ribs with branching**. Shell interior is blue to brown with a septum (small shelf or deck) at the small end of the triangular shell.

SIZE: To 1 3/4" (4.5 cm) long.

HABITAT: On or under rocks and in crevices from the high to mid-intertidal zone.

RANGE: Monterey, California, to Cabo San Lucas, Baja California Sur, México.

NOTES: The branch-ribbed mussel is commonly found in the mid-interdial zone among California mussels (see p. 125), and sometimes in the low intertidal zone. As with most intertidal bivalves, the main predator is the purple star (see p. 197).

CALIFORNIA DATEMUSSEL

Adula californiensis

OTHER NAMES: California pea-pod borer, pea pod borer, *Botula californiensis*.
DESCRIPTION: Shiny chocolate-brown periostracum (shell covering). Shell is elongated with fine hair-like threads at basal end.
SIZE: To 2 1/2" (6 cm) long.
HABITAT: In clay and soft rock, intertidal zone to depths of 66' (20 m).
RANGE: Graham Island, Queen Charlotte Islands, BC, to Point Loma, San Diego County, California.
NOTES: This mussel's shells are very fragile. It is an uncommon species that can sometimes be found at the same sites as other rock-boring bivalves. Scientists know little of its biology.

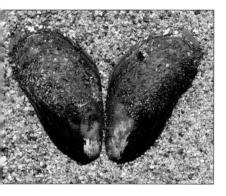

NORTHERN HORSEMUSSEL

Modiolus modiolus

OTHER NAMES: Bearded mussel, horse mussel, *Volsella modiolus*.
DESCRIPTION: Violet-colored rhomboidal shell with a brown, normally **hairy periostracum (covering)**.
SIZE: To 7" (18 cm) long.
HABITAT: Often partially buried in gravel or rocks, intertidal zone to depths of 660' (200 m).
RANGE: Bering Sea to Monterey, California.
NOTES: Unlike most mussels, northern horsemussels can burrow, and often cluster together in groups. Their byssus (cluster of thread-like fibers) attaches to the surrounding rock and to their buddies, aiding them to stay in place. This mussel is circumpolar in distribution. It is not considered a gourmet item.

Straight Horsemussel
Modiolus rectus

OTHER NAMES: Fan horsemussel, fan-shaped horse mussel, fat horse mussel, giant horse mussel; formerly *Volsella flabellatus, V. recta.*
DESCRIPTION: Shell is white to bluish white and covered with a glossy dark brown periostracum (covering). **Shell is narrow and elongated**, with a hinge near the narrow (anterior) end.
SIZE: To 9" (22 cm) long.
HABITAT: Buried in muddy sand or gravel from the low intertidal zone to depths of 150' (46 m).
RANGE: Tow Hill, Queen Charlotte Islands, BC, to Peru.
NOTES: This horsemussel is found in quiet bays or lagoons, or in offshore areas. It is a solitary species that lays buried with only its posterior tip left uncovered. Byssal threads (thread-like fibers) keep this bivalve anchored in the sand or mud.

• •

Olympia Oyster
Ostrea conchaphila

OTHER NAMES: Native Pacific oyster, California oyster, *Ostrea lurida.*
DESCRIPTION: Small gray oyster with **round to elongated shells**.
SIZE: To 3 1/2" (8.8 cm) in diameter.
HABITAT: At a variety of sites, including tidepools and estuaries with mud or gravel flats, intertidal zone to depths of 165' (50 m). Solid objects such as rocks are required for this species to attach.
RANGE: Sitka, Alaska, to Panama.
NOTES: This native species is capable of changing its sex during spawning season. Like several other species, it is called a protandrous hermaphrodite: it is never totally male or female, but alternates between male and female phases. The female holds the young inside her mantle until their shells are developed. While the young develop, the male gonads begin the process of producing sperm. And so the phases alternate. Because of pollution, overharvesting and the species' slow growth (it requires 4 to 6 years to mature), its populations have declined sharply. A similar species, the introduced Pacific oyster (p. 130), is a larger oyster whose shells have fluted edges.

PACIFIC OYSTER

Crassostrea gigas

OTHER NAMES: Japanese oyster, giant Pacific oyster.
DESCRIPTION: Gray to white shells with purple to black new growth. **Irregular shape with fluted exterior edge**. Lower shell is cup-shaped and larger than top shell.
SIZE: **To 12" (30 cm) long**.
HABITAT: Intertidal zone to depths of 20' (6 m).
RANGE: Prince William Sound, Alaska, to Newport Bay, California

NOTES: The Pacific oyster was introduced to BC and Washington in 1922. It can be harvested commercially after only 2 to 4 years but is known to live longer than 20 years. If you are harvesting this oyster, leave the shells on the beach. They provide attachment sites for new generations of oysters. Harvesters should possess a license and be aware of bag limits and closures, especially for red tide (PSP) (see p. 13).

ATLANTIC OYSTER

Crassostrea virginica

OTHER NAMES: Eastern oyster, American oyster, Virginia oyster, blue point oyster, commercial oyster.
DESCRIPTION: Shell exterior is tan to purple; **interior is white or yellow with a black or purple muscle scar**. The irregular shells widen from a **narrow beak to a broad, flat upper valve** and a cupped lower valve.
SIZE: To 8" (20 cm) long.
HABITAT: In estuaries from the low intertidal zone to water 40' (12 m) deep.
RANGE: Boundary Bay, BC, to Tomales Bay, California.
NOTES: The Atlantic oyster, valued for its excellent flavour, was introduced from the east coast around 1870. Today few remnant populations persist along the Pacific coast, including Boundary Bay, BC, and a commercial operation at Tomales Bay, California. It is believed that the Atlantic oyster can live to 20 years, but individuals are large enough to be marketed at about age 3.

Hemphill Fileclam

Limaria hemphilli

OTHER NAMES: File shell, Hemphill's lima; formerly *Lima hemphilli*.

DESCRIPTION: Shell is white, oblique in shape, and widest at the ventral end. The soft interior is orange to pink in color. The shell has fine radial ribs and a gape at both the anterior and posterior ends.

SIZE: To 3/4" (2 cm) high.

HABITAT: Under rocks and on wharf pilings from the low intertidal zone to water 330' (100 m).

RANGE: Monterey, California, to Kobe Beach, Panama.

NOTES: The Hemphill fileclam makes a "nest" of rubble from which it can swim away. In order to swim, it closes its valves (shells) rapidly, ejecting water on either side of the hinge. The margin contains sticky tentacles that can be shed when the animal is attacked. Eyes are also positioned between the valves at the margin (edge).

Giant Rock Scallop

Crassadoma gigantea

OTHER NAMES: Purple-hinged rock scallop, *Hinnites giganteus*, *Hinnites multirugosus*.

DESCRIPTION: Shell color varies from brown to green and is often obscured by myriad encrusting species. Round, thick shells with deep purple color on inside hinges. When shells are open, mantle is visible—usually bright orange and lined with many tiny blue eyes.

SIZE: To 10" (25 cm) in diameter.

HABITAT: In rocky areas, low intertidal zone to depths of 150' (45 m).

RANGE: Prince William Sound, Alaska, to Bahía Magdalena, Baja California Sur, México.

NOTES: This is a free-swimming species until it reaches approximately 1" (2.5 cm) in diameter. At that time it usually attaches to a rock or shell, where it remains for the rest of its life. Older individuals are often found with encrusting algae or boring sponges growing on the shells. It can live as long as 50 years. This species is a gourmet item. If you harvest it, ensure that you are aware of area closures, bag limits and protected areas where harvesting is not allowed.

KELP SCALLOP
Leptopecten latiauratus

OTHER NAMES: Wide-eared scallop, broad-eared pecten, kelp-weed scallop; formerly *Pecten monotimeris*.
DESCRIPTION: Shell color varies widely from orange to

The kelp scallop is found in a wide array of colors.

yellowish brown or brown. Brown and white zigzag or V-shaped lines are prominent on most shells but absent on orange-colored individuals. Shells are thin, slightly convex with 9–20 ribs radiating from the hinge area. The front ear is shorter than the hind ear.
SIZE: To 2" (5 cm) in diameter.
HABITAT: Attached to rock, kelp and similar objects from the low intertidal zone to water 825' (250 m) deep.
RANGE: Point Reyes, Marin County, California, to Cabo San Lucas, Baja California Sur and Gulf of California, México.
NOTES: The empty shells of the kelp scallop are sometimes found washed up on shore. These small scallops are normally observed living on pilings, eel-grass and algae. Offshore, they can be found living on shells, tube worms or pebbles.

PACIFIC CALICO SCALLOP
Argopecten ventricosus

OTHER NAMES: Speckled scallop, pecten, fan shell; formerly *Plagioctenium circularis, Aequipecten aequisulcatus, A. circularis*.
DESCRIPTION: Shell exterior varies from **cream** to orange **with dark reddish or orange blotches**. Valves are **deeply convex** with 19–22 distinct radial ribs. Ears are almost equal in size.
SIZE: To $3^{3/4}$" (9.5 cm) in diameter.
HABITAT: In sandy or muddy waters 10–150' (3–50 m) deep.
RANGE: Santa Barbara, California, to Peru
NOTES: The Pacific calico scallop is an active swimmer with tiny, highly developed eyes. It is also highly sought-after in the culinary world. The empty shells of this beautiful bivalve are commonly found washed up on sandy beaches in California. Live animals are seen in sheltered lagoons, estuaries or bays, such as Newport Bay. This species can temporarily attach to rocks by producing a few byssal threads. Individuals in the northern part of the range are less convex and not as brightly colored as southern individuals. The 2 forms were formerly believed to be distinct species.

San Diego Scallop
Euvola diegensis

OTHER NAMES: Butterfly scallop; formerly *Pecten diegensis*.
DESCRIPTION: Right valve (shell) yellowish and **left valve reddish to orange with a white central area**. Left valve is flat or slightly convex; right valve is convex with 21 or 22 convex topped ribs. Wing-like projections are present at the umbo (apex).
SIZE: Normally to $3^{1/2}$" (9 cm) in diameter, occasionally to 6" (15 cm).
HABITAT: In water 30–450' (9–135 m) deep.
RANGE: Bodega Bay, California, to Cabo San Lucas, Baja California Sur, México.

NOTES: The empty shells of this scallop are sometimes found on sandy beaches after storms. These shells were treasured by native Americans, who used them for decoration.

Pearly Jingle
Anomia peruviana

OTHER NAMES: Peruvian jingle shell, rock oyster, jingle shell, southern jingle.
DESCRIPTION: Thin shell with a smooth exterior that varies from **translucent orange** to yellowish green to copper, with a **subdued luster**. A large notch is prominent on the flat lower valve. **3 small muscle scars** can be seen in the central area of the upper valve.
SIZE: To 2" (5 cm) in diameter.
HABITAT: Attached to rock or shells from the low intertidal zone to water 150' (46 m) deep.
RANGE: Monterey, California, to Peru.
NOTES: The upper shells, which are considerably larger than the lower ones, are often found washed up on shore. The distinctive coppery-colored shells are especially easy to identify.

GREEN FALSE-JINGLE
Pododesmus macroschisma

OTHER NAMES: Rock oyster, jingle shell, abalone jingle shell, false Pacific jingle shell, pearly monia; formerly *Pododesmus cepio*; occasionally misspelled as *P. machrochisma*.
DESCRIPTION: Gray-white upper shell with radiating light lines and iridescent green interior. Lower shell is smaller and flat with a large hole in the middle. **2 muscle scars** are located on the interior of the upper valve.
SIZE: To $5^{1/8}$" (13 cm) long.
HABITAT: Attached to solid objects, low intertidal zone to depths of 300' (90 m).
RANGE: Alaska to Gulf of California.
NOTES: This species is edible, though it is not an abundant intertidal species. Its bright orange mantle is sometimes visible when the shells are open. The distinctive shells are often washed up on beaches. Its name likely originates from the sound the shells make jingling in a beachcomber's pocket.

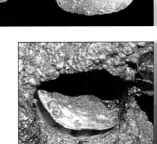

CALIFORNIA LUCINE
Epilucina californica

OTHER NAME: Formerly *Lucina californica*.
DESCRIPTION: Shell is whitish to yellowish and rounded in shape. Beaks are centrally placed and the external ligament is located in a deep pit.
SIZE: To $1^{1/2}$" (4 cm) long.
HABITAT: In sand and gravel from the low intertidal zone to water 265' (80 m) deep.
RANGE: Crescent City, California, to Rocas Alijos, Baja California Sur, México.
NOTES: The California lucine is found on exposed shores. Very little is known about the biology of this bivalve.

Secret Jewelbox
Chama arcana

OTHER NAMES: Clear jewel box, rock oyster, frilled hoof shell, right-hand chama, agate chama; often incorrectly listed as *Chama pellucida*, a species from South America.
DESCRIPTION: Color of the nearly circular shells varies from opaque to translucent white. Right valve (shell) is often tinged with pink or orange, slightly convex with several concentric ridges bearing leafy projections. Left valve is cupped and attached to sub-

Shell projections have been worn off.

strate. **Beak is coiled to the right (clockwise)** when viewed from above.
SIZE: To $2^{1/2}$" (6 cm) in diameter.
HABITAT: On rocks and pilings from the low intertidal zone to 260' (78 m).
RANGE: Waddell Beach, Santa Cruz County, California, to Bahía Magdalena, Baja California Sur, México.
NOTES: The secret jewelbox is one of two species of jewelbox or chama found on the North American coast. The hinged empty shells of these distinctive bivalves truly look like jewel boxes. Subtidal populations of the secret jewelbox pile up into thick crusts and can reach a population density of more than 400 per .1 sq meter. Despite its small size, it is eaten by sea otters and occasionally man.

Pacific Left-Handed Jewelbox
Pseudochama exogyra

OTHER NAMES: Left-handed jewel box, Pacific jewel box, reversed chama, California reversed chama, rock oyster, frilled hoof shell.
DESCRIPTION: Color of the nearly circular shells varies from opaque to translucent white, often tinged with reddish brown. **Beak is coiled to the left (counter-clockwise)** when viewed from above. The frills on the shells are often badly worn.
SIZE: To 2" (5 cm) in diameter.
HABITAT: On exposed rocky reefs in the mid-intertidal zone to 80' (25 m).
RANGE: Oregon to Cabo San Lucas, Baja California Sur, México.
NOTES: The Pacific left-handed jewelbox is similar to the secret jewelbox except that the opposite valve is attached to the substrate; as a result the direction of its coil is reversed. The 2 species are often found side by side on the same substrate.

MOLLUSKS AND LAMPSHELLS

KELLYCLAM
Kellia suborbicularis

OTHER NAMES: North Atlantic kellia, suborbicular kellyclam, smooth kelly clam; formerly *K. comandorica*; includes *K. laperousii*.
DESCRIPTION: Shell is white with a thin, yellowish periostracum (covering). The shell is **globular in shape**.
SIZE: To 1" (2.5 cm) long.
HABITAT: In crevices from the low intertidal zone to water 65' (20 m) deep.
RANGE: Prince William Sound, Alaska, to Peru.
NOTES: The kellyclam broods its young internally before releasing the veligers (tiny young). The free-swimming young eventually settle down and attach themselves to rocks with a byssus (threads). This clam is often found inside empty pholad (boring clam) holes as well as discarded bottles. It is circumboreal in its distribution.

LITTLE HEART CLAM
Glans carpenteri

OTHER NAMES: Carpenter's carditid; formerly *G. subquadrata*, *G. minuscula*.
DESCRIPTION: Oblong shell is yellowish white with reddish brown spots. Beaded radial ribs extend from the hinge area.
SIZE: To $5/8$" (1.5 cm) long.
HABITAT: On the undersides of rocks from the low intertidal zone to water 330' (100 m) deep.
RANGE: Frederick Island, BC, to Punta Rompiente, Baja California Sur, México.
NOTES: The little heart clam produces eggs that are released into the mantle cavity. Here they are fertilized and brooded until they grow large enough to go out on their own. A young clam has a shell and sometimes starts out attaching to its parent's shell. It is able to crawl about with its large foot, and to reattach elsewhere with byssal threads.

Nuttall's Cockle
Clinocardium nuttallii

OTHER NAMES: Cockle, basket cockle, heart cockle, *Cardium corbis*.
DESCRIPTION: Shell color varies from light brown with mottling in young individuals to much darker in older specimens. Shells are roughly oval with **heart-shaped cross-section**, prominent radiating ribs on exterior.
SIZE: Occasionally to $5^{1/2}$" (14 cm) long.
HABITAT: On both sand and mud beaches, mid-intertidal zone to subtidal depth of 80' (24 m).
RANGE: Bering Sea, Alaska, to San Diego, California.

NOTES: This cockle has been known to live 16 years. Breeding may take place at 2 years. Since its siphons are very short, this species is usually found on or near the surface of tidal flats. Predatory sea stars such as the giant pink star (see p. 196) and the sunflower star (p. 197) feed on this clam. The Nuttall's cockle uses its long foot in a remarkable thrusting motion—pole-vault style—to push off from its enemies.

• •

Pacific Eggcockle
Laevicardium substriatum

OTHER NAMES: Egg cockle, common Pacific egg cockle, little egg cockle; formerly *Cardium pedernalense*.
DESCRIPTION: Ovate shell is pale yellow to gray mottled with brown, smooth and often sculpted with fine radial ribs. Interior is mottled with purplish brown.
SIZE: To $1^{1/2}$" (4 cm) long.
HABITAT: On sandy bottoms from the low intertidal zone to water 130' (40 m) deep.
RANGE: Mugu Lagoon, California, to Bahía Santa Maria, Baja California Sur, México.
NOTES: This small, common clam buries itself shallowly in the sand or mud of bays and estuaries. Some marine biologists believe that it is the same as another species, *L. elenense*, which lives as far south as Peru.

SPINY COCKLE
Trachycardium quadragenarium

OTHER NAMES: Forty-ribbed cockle, giant Pacific cockle, spiny pricklycockle; sometimes misspelled *T. quadrigenarium*.
DESCRIPTION: Shell colour varies from gray to brown buff or yellow. Shell is sculpted with 41–44 strong radial ribs with **triangular spines near the edge**.
SIZE: To 6" (15 cm) long.
HABITAT: In sandy mud from low intertidal zone to water 165' (50 m) deep.
RANGE: Monterey, California, to Punta Rompiente, Baja California Sur, México.
NOTES: The spiny cockle is a large species that buries itself just under the surface of the sand. Live specimens are not common enough to be used as a food. Their empty shells are sometimes found washed up on shore.

PACIFIC GAPER
Tresus nuttallii

OTHER NAMES: Summer clam, rubberneck clam, big-neck clam, horse clam, otter-shell clam, great Washington clam, *Schizothoerus nuttalli*.
DESCRIPTION: White to yellowish shells, brown to black periostracum (shell covering), which peels easily. Noticeably **oval-elongated shells**. A gape (opening) is present between the valves for the siphon.
SIZE: To 4 lbs (1.8 kg), to 8" (20 cm) long.
HABITAT: In sandy areas, from the mid-intertidal zone to depths of 265' (80 m). Can burrow as deep as 36" (90 cm) into the sand.
RANGE: Kodiak Island, Alaska, to Bahía Magdalena, Baja California Sur, México.
NOTES: This clam is noted for sporadically spurting jets of water approximately 3' (1 m) into the air. It is believed to be a summer spawner. Native people of the Northwest Coast dried the siphons of this species for winter food.

A siphon "show."

Hemphill Surfclam

Mactromeris hemphillii

Other Names: Hemphill's surfclam, Hemphill's dish clam; formerly *Spisula hemphillii*.
Description: Yellowish white shell with a thin yellowish brown periostracum (covering). Fine concentric growth rings are sculpted on the thin, smooth shells. **The front of the shell's upper edge is concave**; the hind edge is convex. Valves do not gape. The spoon-shaped chondrophore projects ventrally.
Size: To 6" (15 cm) long.
Habitat: In sand or sandy mud from the very low intertidal zone to water 150' (46 m) deep.
Range: Santa Barbara, California, to Punta Pequeña, Baja California Sur, México.
Notes: The Hemphill surfclam is found in bays, estuaries, sloughs and sheltered coastal areas with fine sand or sandy mud as a base. Although this species is not often taken by clammers, its flesh is quite tasty.

Rosy Jackknife Clam

Solen rostriformis

Other Names: Rosy razor clam; formerly *Solen rosaceus*.
Description: Shell exterior is rosy or pink to gray-white with a thin, transparent, varnish-like periostracum (covering). Shells are very fragile, elongated and oblong. The hinge is at the extreme anterior end.
Size: To 3" (7.5 cm) long.
Habitat: On sand in protected bays in the low intertidal zone to 165' (50 m).
Range: Santa Barbara, California, to Gulf of California.
Notes: The rosy jackknife clam burrows to 12" (30 cm) deep in a permanent burrow in which it can move up and down quickly. Food consists of plankton and suspended organic detritus that is brought in through the animal's siphon. This is an uncommon species that is sometimes dug up by people clamming for other species. Its empty shells are occasionally washed high up on a sandy beach; these light shells are fragile and are often damaged by the waves that tossed them ashore.

SICKLE JACKKNIFE CLAM
Solen sicarius

OTHER NAMES: Blunt razor-clam, sickle razor clam; formerly *Solen perrini*.
DESCRIPTION: Shells are smooth and shiny with greenish yellow to brown periostracum (covering), and nearly rectangular in shape.
SIZE: To 5" (12.5 cm) long.
HABITAT: In sheltered sand-mud areas from the low intertidal to 180' (55 m).
RANGE: Queen Charlotte Islands, BC, to Bahía San Quintin, Baja California, México.
NOTES: The sickle jackknife clam is often found near eel-grass beds, where it builds a permanent burrow to 14" (35 cm) deep. In this burrow it can travel very rapidly. This species has also been known to bury itself completely in 30 seconds and to leap "several centimeters" from the sand. It swims by jetting water from its siphons or the area around its foot. The latter method is likely useful in its rapid digging technique.

PACIFIC RAZOR CLAM
Siliqua patula

OTHER NAME: Razor clam.
DESCRIPTION: Yellow to yellowish brown shells with glossy, smooth finish; flattened oval in shape. Foot and siphon are buff-colored.
SIZE: To 6" (15 cm) long, $2^{3/8}$" (6 cm) wide.
HABITAT: On surf-swept sandy beaches, low intertidal zone to water 180' (55 m) deep. Digs to depths of 24" (60 cm).
RANGE: Cook Inlet, Alaska, to Morro Bay, California.
NOTES: This clam is very active and has been observed to bury itself completely in sand in less than 7 seconds. To do so, the clam pushes its foot deeper into the sand while its fluids are displaced. The tip of the foot then expands, forming an anchor. Then, as the foot contracts, the animal draws deeper into the sand. Individuals have been known to live 18 years. Pacific razor clams were traditionally gathered in May and June, at low spring tides, by Native people. Today they are collected by both recreational and commercial fishermen. If you harvest this species, be sure that the area is safe from red tide or PSP (see p. 13).

CALIFORNIA JACKKNIFE CLAM
Tagelus californianus

OTHER NAMES: Californian tagelus.
DESCRIPTION: Shell is white to gray with a dull brown periostracum (covering). The valves (shells) have near-parallel margins and a central hinge.
SIZE: To 5" (13 cm) long.
HABITAT: In sandy mud flats of bays in the low intertidal zone.
RANGE: Humboldt Bay, California, to Costa Rica.
NOTES: The California jackknife clam is a common inhabitant of mud or sand-mud of estuaries, bays and sloughs. Here it digs to a depth of 20" (51 cm) but usually remains near the mouth of its burrow, utilizing 2 separate holes for its siphons. The closely related Pacific razor clam (see p.140) has fused siphons and thus utilizes a single hole. The California jackknife clam is sometimes used as bait for fishing.

BALTIC MACOMA
Macoma balthica

OTHER NAMES: Tiny pink clam, *Macoma inconspicua*.
DESCRIPTION: Oval shells are often pink but can also be blue, orange or yellow.
SIZE: To 1 1/2" (4 cm) long.
HABITAT: In areas with mixed mud and sand, mud flats and eel-grass beds, intertidal zone to depths of 130' (39 m).
RANGE: Beaufort Sea to San Francisco Bay, California.
NOTES: This common species is often plentiful in muddy areas. It also lives in northern Atlantic waters. Members of the genus *Macoma* have 2 separate siphons, one for water to enter and the other for it to leave. The siphons extend to the surface for the animal to feed, breathe and expel waste.

MOLLUSKS AND LAMPSHELLS

INDENTED MACOMA
Macoma indentata
OTHER NAME: Formerly *M. rickettsi*.
DESCRIPTION: Shells are **white and glossy** with a very thin, skin-like periostracum (covering). The valves are thin and flexed (curved or bent), with an **indentation in the edge near the pointed posterior end**.
SIZE: To 4" (10 cm) long.
HABITAT: In sand or silt from the low intertidal zone of bays to offshore waters 300' (91 m) deep.
RANGE: Little River, Humboldt County, California, to Isla Santa Margarita, Baja California Sur, México.
NOTES: Empty shells of the indented macoma are often found washed up on shore. Juvenile gaper pea crabs (see p. 186) have been found in the cavities of these medium-sized clams, giving them temporary homes until they find larger clam hosts.

BENT-NOSE MACOMA
Macoma nasuta
OTHER NAME: Bent-nosed clam.
DESCRIPTION: Shell has thin, brown, wrinkled periostracum (covering). The clam gets its name from its shells, which are noticeably curved to the right.
SIZE: Normally to 2" (5 cm) long but can reach $4^{1/2}$" (11 cm).
HABITAT: In protected areas with mud and sand, intertidal zone to water 165' (50 m) deep. Can burrow 4–6" (10–15 cm) below the surface.
RANGE: Montague Island, Cook Inlet, Alaska, to Punta Rompiente, Baja California Sur, México.
NOTES: This clam typically lies flat with its "nose" pointed upward, which allows the siphons to reach the surface. Shells from this species have been found in many middens in the Northwest Coast area. Lewis's moonsnail (see p. 94) is known to be a predator of this clam.

White-Sand Macoma

Macoma secta

OTHER NAME: White sand clam.
DESCRIPTION: Oval shells are white with gray or brown paper-like periostracum (covering). The left shell is flatter and often more worn than the right.
SIZE: To 4" (10 cm) long.
HABITAT: On sandy shores from the low intertidal zone to water 165' (50 m) deep. The clam buries to 18" (45 cm) below the surface.
RANGE: Queen Charlotte Islands, BC, to Bahía Magdalena, Baja California Sur, México.
NOTES: The white-sand macoma favours sites with fine sand or sandy mud. It buries itself so deep in the sand that few predators can reach it. There it lies on its flatter left side, extending 2 separate white siphons to the surface up to 2" (5 cm) apart. The inhalant siphon acts like a vacuum cleaner, bringing organic detritus into the gut along with great quantities of sand. The second (exhalant) siphon returns all wastes to the sand's surface. Pea crabs are often found inside this clam.

Bodega Tellin

Tellina bodegensis

OTHER NAME: Bodega clam.
DESCRIPTION: White elongated shells have polished surfaces and very fine concentric lines on exterior. Interior is white, often with a tint of yellow or pink.
SIZE: To $2^{3/8}$" (6 cm) long.
HABITAT: On exposed beaches, intertidal zone to depths of 300' (91 m).
RANGE: Sitka, Alaska, to Bahía Magdalena, Baja California Sur, México.
NOTES: This is an uncommon species but its distinctive shells are sometimes found washed up on the beach. The bodega tellin has 2 separate siphons extending to the surface. It is reportedly an excellent tasting clam, but is too scarce to be harvested regularly.

MOLLUSKS AND LAMPSHELLS

BEAN CLAM
Donax gouldii

OTHER NAMES: Conquina, Gould beanclam, *D. gouldi*.
DESCRIPTION: Color varies from cream to buff, orange or blue with dark rays often present. Shell is heavy and wedge-shaped with highly polished periostracum (covering) and indistinct radiating grooves.
SIZE: To 1" (2.5 cm) long.
HABITAT: On exposed sandy shores from the mid-intertidal zone to 100' (30 m).
RANGE: Pismo Beach, California, to Arroyo del Conejo, Baja California Sur, México.
NOTES: Populations fluctuate wildly. In fact, this species may be found in numbers up to 16,700 per square yard (20,000 per m^2). Individuals bury themselves in the sand in a band 7–16' (2–5 m) wide for stretches up to 5 miles (8 km). However, such a population explosion may occur only once every 14 years. The clam hydroid (see p. 21) is commonly found attached to the exposed posterior end of the bean clam. The wedge clam *Donax californicus*, a similar clam with a triangular shape, is found from Santa Barbara to Baja California, México.

CALIFORNIA MAHOGANY-CLAM
Nuttallia nuttallii

OTHER NAMES: Purple clam, Nuttall's mahogany clam; formerly *Sanguinolaria nuttallii*, *S. orcutti*.
DESCRIPTION: Shell has purplish rays covered with a dark brown, shiny skin-like periostracum (covering). Shell is oval with a long posterior; right valve is nearly flat, left valve is convex. Interior is white to light purple.
SIZE: To 6" (15 cm) long.
HABITAT: In sand or gravel in the low intertidal zone.
RANGE: Bodega Harbor, California, to Bahía Magdalena, Baja California Sur, México.
NOTES: This large clam is found in bays and protected areas. It normally rests on its right side, buried 12–16" (30–40 cm) below the surface. It extends its 2 siphons to the surface and feeds on microscopic foods suspended in the water.

California Sunset Clam

Gari californica

OTHER NAMES: Sunset clam.
DESCRIPTION: Shell is white, **often with faint pink rays** and a thin periostracum (covering). Interior is white and shiny. **Shell is oval** and smooth.
SIZE: To 4 1/4" (10.6 cm) long.
HABITAT: In sand or gravel from the low intertidal zone to water 165' (50 m) deep.
RANGE: Aleutian Islands, Alaska, to Bahía Magdalena, Baja California Sur, México.
NOTES: This clam is found both on the open coast and in entrances to bays. Its colourful shells are found frequently on sandy beaches. The clam burrows to 8" (20 cm) deep where it extends 2 separate siphons to the surface of the sand or gravel.

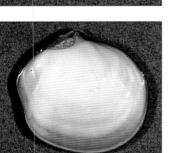

Clipped Semele

Semele decisa

OTHER NAME: Bark semele.
DESCRIPTION: Shell is yellowish white with a thin brown periostracum (covering). Interior is white with reddish purple near the hinge. The heavy shells vary from nearly circular to ovate, with several coarse, irregular rounded concentric ridges.
SIZE: To 4 3/4" (12 cm) long.
HABITAT: In coarse sand or gravel from the low intertidal zone to water 100' (30 m) deep.
RANGE: Point Arguello, California, to Bahía Magdalena, Baja California Sur, México.
NOTES: Empty shells of the clipped semele are often found washed up on shore. Holdfasts of various seaweeds are often attached to the shells, causing these clams to be caught in the turbulent waters of a storm and tossed on shore.

ROCK SEMELE
Semele rupicola

OTHER NAME: Semele-of-the-rocks.
DESCRIPTION: Shell is yellowish white; interior is white with an orange and red margin. Shell is heavy and oval with irregular concentric ridges.
SIZE: Normally to 1 3/4" (4.5 cm), but can reach 2 7/8" (7.1 cm) long.
HABITAT: In crevices from the low intertidal zone to water 180' (55 m) deep.
RANGE: Farallon Island, California, to Bahía Magdalena, Baja California Sur, México.
NOTES: The rock semele is a nestling species. It may have distorted valves. This species has been also found in colonies of California mussel (see p. 125) and inside discarded bottles. Little is known about its biology.

WHITE VENUS
Amiantis callosa

OTHER NAMES: Pacific white venus, sea cockle, white amiantis.
DESCRIPTION: Shell is heavy with a white exterior and a thin, transparent, yellowish periostracum (covering). Shell has pronounced regular, rounded concentric ridges.
SIZE: To 4 1/2" (11.4 cm) long.
HABITAT: On exposed sandy beaches near the entrances to bays, from the low intertidal zone to water 65' (20 m) deep.
RANGE: Santa Barbara, California, to Bahía Santa Maria, Baja California Sur, México.
NOTES: The white venus is rated as delicious by gourmets, but it is not often gathered because it is uncommon. Clammers are more likely to find its empty shells washed up on shore after a storm. This clam broods its eggs in its mantle cavity.

BANDED VENUS
Chione californiensis

OTHER NAMES: Banded chione, Californian chione, California venus, common California venus; formerly *C. succincta, C. durhami, Venus succincta.*

DESCRIPTION: Thick, heavy shell is gray overall, often tinged with brown in the front, and sculptured with **very prominent, well-spaced, raised concentric ridges** and weaker, flattened radial ribs. The concentric ridges are raised in distinctive thin veins.

SIZE: To 3" (7.5 cm) long.

HABITAT: In sand flats of bays from the low intertidal zone to 165' (50 m).

RANGE: San Pedro, California, to Panama.

NOTES: The banded venus is a choice, but not abundant, edible clam. Like all of California's *Chiones*, this one is prized for its excellent taste. This clam has experienced years of mass mortality at various bays in Baja California.

WAVY VENUS
Chione undatella

OTHER NAMES: Wavy chione, frilled California venus, frilled venus; formerly *C. taberi, Venus neglecta.*

DESCRIPTION: Exterior of the heavy grayish white shell is often adorned with a variety of brown patterns or blotches. Interior is white, occasionally with a purple tinge. Shell is sculptured with **prominent radiating ribs and closely spaced concentric ridges**.

SIZE: To 2 1/2" (6 cm) long.

HABITAT: On mud flats and back bays from the intertidal zone to water 130' (40 m) deep.

RANGE: San Pedro, Los Angeles County, California, to Peru.

NOTES: The wavy venus is a very common species with an extensive history of both commercial and sport fisheries in southern California. Its shell is more elongated than that of the banded venus (see above) but little is known about the natural history of this or any other *Chione* species.

SMOOTH VENUS
Chione fluctifraga

OTHER NAMES: Smooth chione, smooth California venus, smooth Pacific venus; formerly *Venus gibbosula*.
DESCRIPTION: Valves are yellowish white overall with a semi-gloss exterior; **interior** is usually tinged **with a purple edge**. The yellowish periostracum (covering) is often eroded, revealing a somewhat chalky shell. **Exterior is smooth** overall with a subdued sculpture of **broad radial ribs and closely spaced low concentric ridges**.
SIZE: To 3 1/2" (8.6 cm) long.
HABITAT: On mud and sandy shores from the intertidal zone to water 80' (25 m) deep.
RANGE: Coos Bay, Oregon, to Isla San Martin, Baja California, México.
NOTES: The smooth venus is the largest *Chione* found in California. It is a clammer's favorite for making chowders.

ROCK VENUS
Irusella lamellifera

OTHER NAMES: Californian irus venus, lamellar venus, ribbed clam; formerly *Irus lamellifer, Venerupis lamellifera*.
DESCRIPTION: Shell is chalky white with approximately 12 **prominent raised, blade-like concentric ridges**. Shell is normally oblong but occasionally nearly round.
SIZE: To 2" (5 cm) long.
HABITAT: In rock holes from the low intertidal zone to water 330' (100 m) deep.
RANGE: Coos Bay, Oregon, to Isla San Martin, Baja California, México.
NOTES: The rock venus is a nestling clam that resides in the empty burrows of various rock-boring clams. The prominent ridges on the shell help the clam to brace itself in its environment.

Japanese Littleneck
Venerupis philippinarum

Other names: Manila clam, Japanese littleneck clam; formerly *Tapes philippinarum, Tapes japonica*.
Description: Light brown to gray shells, often with patterns of streaks and blotches of darker browns. **Oval-elongated shells** with both radiating and concentric raised lines and **smooth inside margin**. Interior is often touched with purple.
Size: To 3" (7.5 cm) long.
Habitat: In muddy gravel, high intertidal to shallow subtidal zone. Burrows to depth of approximately 4" (10 cm).
Range: Southern Queen Charlotte Islands, BC, to Elkhorn Slough, California.
Notes: This species was accidentally introduced to North America with the seed of Pacific oysters. Mass winter mortalities have been noted, which may indicate that this clam cannot withstand extreme temperature changes. Japanese littlenecks require only 2 years to reach legal harvestable size of 2" (5.1 cm). If you plan to harvest them, check with officials to ensure that the area is free from pollution and red tide.

Rough-sided Littleneck
Protothaca laciniata

Other names: Rough cockle, bay cockle.
Description: Shell is buff to orange-brown and may have darker stains. Shell is heavy with a **rasp-like exterior**. Concentric ridges and radiating ribs are equally prominent.
Size: To $3^{1/4}$" (8 cm) long.
Habitat: In sheltered shores from the intertidal zone to water 16' (5 m) deep.
Range: Monterey Bay, California, to Laguna San Ignacio, Baja California, México.
Notes: The rough-sided littleneck is normally found a mere 2–3" (5–7 cm) beneath the surface of the substrate but large specimens have been found as deep as 10" (25 cm). This clam occurs in a wide range of sites, including the firm sandy mud of bays, sloughs and estuaries as well as in semi-protected coastal waters.

PACIFIC LITTLENECK

Protothaca staminea

OTHER NAMES: Rock cockle, rock venus, rock clam, hardshell clam.
DESCRIPTION: White to light brown shells, often with fine irregular brown pattern. Generally **round shells** with both radiating and concentric raised lines. **Inside margin of shell has rough texture**.
SIZE: To 3" (7.5 cm) long.
HABITAT: On sand and gravel beaches, mid-intertidal zone to water 33' (10 m) deep. Can burrow to a depth of approximately 4" (10 cm).
RANGE: Aleutian Islands, Alaska, to Bahía Santa Maria, Baja California Sur, México.
NOTES: This clam is generally slow-growing, reaching legal harvestable size in 3 to 4 years but living as long as 16 years. Breeding occurs during the summer months, after a maturation period of 3 to 5 years. A similar species, the Japanese littleneck (see p. 149), has more elongated shells. Pacific littlenecks make excellent eating when steamed. But you must have a sport fishing license to harvest them legally, and be sure to check with local officials to ensure the clams are safe to eat in the area you wish to harvest from.

CALIFORNIA BUTTER CLAM

Saxidomus nuttalli

OTHER NAMES: Washington clam, common Washington clam, butter clam.
DESCRIPTION: Shell exterior is white to tan in color; interior is a lustrous white, tinged with purple at the siphon end. The shells are heavy and oval with **many heavy concentric rings**.
SIZE: To 4 3/4" (12 cm) long.
HABITAT: In sand, mud or sandy mud from the low intertidal zone to water 150' (46 m) deep.
RANGE: Humboldt Bay, California, to San Quintin Bay, Baja California, México.
NOTES: The California butter clam burrows deep from 12–18" (30–45 cm) into the sand or similar substrate. Another related northern species, the butter clam (*Saxidomus gigantea*), is found as far south as central California. The butter clam lacks the distinctive concentric rings found on the California butter clam.

Pismo Clam
Tivela stultorum

OTHER NAME: Formerly *Tivela scarificata*.
DESCRIPTION: Shell is light tan to brown, often with brownish radiating bands. **Shell is triangular, thick and heavy**, with a smooth, varnish-like periostracum (covering).
SIZE: To 6" (15 cm) long.
HABITAT: On sandy beaches exposed to the surf from the low intertidal zone to water 80' (24 m) deep.
RANGE: Half Moon Bay, San Mateo County, California, to Bahía Magdalena, Baja California Sur, México.
NOTES: The pismo clam is a tasty bivalve whose numbers are closely monitored for limited harvesting. Contact the California Department of Fish and Game to obtain current limits and sizes for harvest. Years ago this clam was harvested in numbers as high as 50,000 daily. As a result, its populations fell dramatically and required protection; they have since recovered. This large clam has short siphons and so it lives near the surface of the sand. It can be found at several sites, including Pismo Beach—a site named after this clam or vice versa.

California Softshell-Clam
Cryptomya californica

OTHER NAMES: False mya, California glass mya.
DESCRIPTION: White or off-white oval shells with brown periostracum (skin-like covering) and a series of concentric rings.
SIZE: To 1 1/4" (3.2 cm) long.
HABITAT: In sand and mud in the intertidal zone, to water 265' (80 m) deep. Burrows to depths of 20" (50 cm).
RANGE: Montague Island, Alaska, to Peru.
NOTES: The siphons of this small clam are only 1/16" (1 mm) long, unusually short for a clam. This species lives deeper in the sand than its siphons would normally allow by living near the burrows of the bay ghost shrimp (p. 171) and using its short siphon to feed from the water circulating there.

SOFTSHELL-CLAM

Mya arenaria

OTHER NAMES: Mud clam, soft clam.

DESCRIPTION: White or gray shells with yellow or brown periostracum (covering). Shells are relatively soft (as the clam's name implies) and slightly elongated with many concentric rings. A chondrophore (large spoon-shaped projection) is located at the hinge of one shell.

SIZE: To 4" (10 cm) long.

HABITAT: Buried in sand and mud in estuaries, intertidal zones. Burrows to a depth of approximately 8" (20 cm).

RANGE: Alaska to Elkhorn Slough, California.

NOTES: The softshell-clam burrows in mud and sand in a unique way. It ejects water below its body, pushing sand out of the way as it digs deeper—a slow method of burrowing that is more effective in sand than in mud. This clam also has a special sac off the stomach that holds a food reserve. The shells of this species are not found in the middens of Native people, which may indicate that it was accidentally introduced to North America in the last century.

BORING SOFTSHELL-CLAM

Platyodon cancellatus

OTHER NAMES: Chubby mya, checked softshell-clam, checked borer.

DESCRIPTION: Yellow-brown periostracum (shell covering) overlays rectangular shells. Narrow, bony projection inside one shell near the hinge.

SIZE: To 3" (7.6 cm) long.

HABITAT: Species builds an elongated burrow in soft rock or clay to 5" (13 cm) deep, often with sand covering the entrance. Intertidal zone to subtidal waters 330' (100 m) deep.

RANGE: Queen Charlotte Islands, BC, to San Diego, California.

NOTES: This clam can bore directly into soft rock by shell movement rather than a rotating motion. It does so by alternate contractions of muscles in the shells, which causes the shells to rock. As a result, it can bore holes that closely match the shape of its shells, rather than the round holes of other boring species. It is also reported to have been found boring into low-grade concrete.

Pacific Geoduck
Panope abrupta

OTHER NAMES: King clam, gooeyduck, *Panopea generosa*.
DESCRIPTION: Gray shells with yellow periostracum (covering). Shells are heavy and oblong with concentric rings. Body and immense siphons cannot be completely contained within shells. The tip of the siphon, when viewed from above, is light brown, without tentacles or pads.
SIZE: Has been reported to reach 9" (23 cm) long; most are much smaller. Largest recorded weight is 20 lbs (9 kg).
HABITAT: In protected beaches of gravel, sand and mud, low intertidal zone to subtidal waters 200' (61 m) deep.
RANGE: Arctic Ocean to Panama.

NOTES: This remarkable clam can extend its siphon nearly 3' (1 m) to the surface, and as a result it is not easily collected. Because the siphon is too large to be retracted completely into the shell, the geoduck relies on its deep burrow for protection. The giant pink star (see p. 196) feeds on the siphons of this clam by grasping them with its sticky tube feet. The Pacific geoduck is the largest burrowing clam in the world. It is also extensively harvested commercially. The total number of eggs produced by a single female during one year has been calculated to exceed 50 million; few of these survive, but individuals of this species have been known to live longer than 140 years.

Beaked Piddock

Netastoma rostratum

OTHER NAMES: Rostrate piddock; formerly *Nettastomella rostrata, Netastoma rostrata.*
DESCRIPTION: The delicate white shells include an **elongated tapering extension** at the siphon end.
SIZE: Shell to 3/4" (2 cm) long.
HABITAT: Burrows in shale and dead shells from the low intertidal zone to water 33' (100 m) deep.
RANGE: Barkley Sound, Vancouver Island, BC, to Bahía San Cristóbal, Baja California Sur, México.
NOTES: The beaked piddock burrows into shale and can become especially abundant in areas with soft shale, such as Monterey Bay. The shape of the shell is often somewhat irregular because of this boring activity. This small species is often found in the company of other rock-burrowing clams.

Arctic Hiatella

Hiatella arctica

OTHER NAMES: Arctic rock borer, Gallic saxicave, little gaper, red nose, *Hiatella gallicana.*
DESCRIPTION: Chalky white shells with rough, round-elongated shape. Siphon tips are red or orange.
SIZE: To 2" (5 cm) long.
HABITAT: Intertidal zone to subtidal waters 2,640' (800 m) deep.
RANGE: Alaska to Chile.
NOTES: The Arctic hiatella bores into soft, uniform rock using a rocking motion. It is also an opportunist, often moving into the burrows of other boring clams. Adults can attach directly to rock surfaces with weak byssus threads.

SCALE-SIDED PIDDOCK
Parapholas californica

DESCRIPTION: Shell is grayish white in color, with a pear shape, a **tapered hind end**. **Elongated plates** cover the upper and lower margins and hind end. Interior is white with a chondrophore (spoon-shaped projection).
SIZE: To 6" (15 cm) long.
HABITAT: Bores into hard clay, soft sandstone or shale from the low intertidal zone to water 30' (9 m) deep.
RANGE: Bodega Bay, California, to Punta Pequeña, Baja California Sur, México.
NOTES: The scale-sided piddock bores up to 12" (30 cm) into the rock substrate, where populations can average 50 per square yard (m^2). These bivalves are the major cause of erosion in rock areas such as the Scripps Submarine Canyon in San Diego County. The siphons of this species can be completely retracted into its shells, unlike the siphons of the flat-tip piddock (see below). The scale-sided piddock is also considered by some to be an aphrodisiac.

FLAT-TIP PIDDOCK
Penitella penita

OTHER NAMES: Common piddock; formerly *Pholadidea penita*.
DESCRIPTION: Gray to brown periostracum (shell covering). Hinge is near one end of the 2 shells. Exteriors of wedge-shaped shells are divided into 3 separate sections. In each shell, the central near-triangular section separates the other 2 sections. A triangular accessory plate is present.
SIZE: To 3" (7.6 cm) long.
HABITAT: Buried to 20" (50 cm) deep in soft rock, mud or hard clay, intertidal zone to water 72' (22 m) deep.
RANGE: Chirikof Island, Alaska, to Bahía San Bartolomé, Baja California, México.
NOTES: This piddock lives in burrows with only the siphon exposed to obtain food. The opening in the rock is typically small, since the clam begins to burrow when it is quite small. Burrowing is accomplished by simple mechanical abrasion of the rock by the shell. Since the animal is continually growing, it must continually enlarge its burrow. Once it has reached its full size, its foot is no longer required; the foot degenerates and the shells overgrow it. The similar rough piddock (see p. 156) is a larger species with its exterior divided into 2 sections.

MOLLUSKS AND LAMPSHELLS

ROUGH PIDDOCK

Zirfaea pilsbryi

OTHER NAME: Pilsbry's piddock; sometimes misspelled *Zirphaea pilsbryi*.

DESCRIPTION: White shells with brown periostracum (covering). Shells are roughly oval with rough concentric growth rings, and teeth developed at anterior end. **Shell exterior is divided into 2 separate sections** and gapes at both ends.

SIZE: To 6" (15 cm) long; siphon can extend to almost 12" (30 cm) long and 2" (5.1 cm) in diameter when extended.

HABITAT: Species burrows into shale and clay, intertidal zone to subtidal depths of 412' (125 m).

RANGE: Bering Sea to Bahía Magdalena, Baja California Sur, México.

NOTES: The rough piddock rasps away at clay or soft rock to build its burrow with sharp tooth-like projections at the rear of the shells. As it digs it rotates, carving a circular burrow. This clam is a true "prisoner in its castle": it is embedded in its home until it dies. The rough piddock has been known to live 8 years. The shells of a similar rock-boring species, the flat-tipped piddock (see p. 155), are divided into 3 sections each.

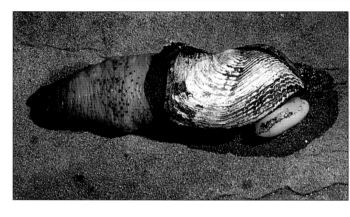

A siphon "show."

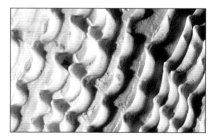

Detail of the rough piddock's cutting surface.

Feathery Shipworm

Bankia setacea

Other names: Teredo, Pacific shipworm.
Description: Superficially a worm-like creature with small shells at the front end and 2 feather-like pallets or appendages at the rear.
Size: To 39" (1 m) long, 1" (2.5 cm) in diameter.
Habitat: Species bores into wood, intertidal zone to subtidal 300' (90 m).
Range: Bering Sea to Gulf of California.

Notes: The young of the shipworm (veligers) look like miniature clams in the first stages of development. Changes occur once they settle on a wood source to begin their "boring" life. They use the teeth on their shells to rasp away at the wood while they burrow. The 2 feather-like appendages that grace the rear can be used to stopper the burrow. Shipworms feed on both wood and plankton as they grow, and they cause considerable damage to untreated wood every year.

Rock Entodesma

Entodesma navicula

Other names: Rock-dwelling entodesma, rock-dwelling clam, northwest ugly clam, *Entodesma saxicola, Agriodesma saxicola*.
Description: Periostracum (shell covering) is brown, and usually cracks when dried out. This cracking often breaks the shell as well. Shell is usually oblong but often shaped by its habitat.
Size: To 6" (15 cm) long.
Habitat: Along shores covered with broken rocks, among rock crevices, intertidal zone to 65' (20 m).
Range: Atka Island, Aleutian Islands, Alaska, to Point Loma, San Diego County, California.

The shell of a young individual.

Notes: The young of this species look much different than adults. A light brown periostracum completely covers the shell of the young, giving it a unique appearance. As this species is sometimes called the northwest ugly clam, you know it is not one of the prettier shells to collect.

MOLLUSKS AND LAMPSHELLS

CALIFORNIA LYONSIA
Lyonsia californica

DESCRIPTION: Shell is translucent and slightly iridescent with a thin, transparent yellow periostracum (covering). Shell is an elongated oval with a straight edge at the posterior end.
SIZE: To 1 3/4" (4.5 cm) long.
HABITAT: In sandy mud from the low intertidal zone to water 33' (100 m) deep.
RANGE: Kodiak Island, Alaska, to Gulf of California.
NOTES: This distinctive bivalve buries itself close to the surface in mud or sandy mud, so that it can reach the surface with its 2 short siphons. It is a simultaneous hermaphrodite, releasing eggs and sperm alternately through the exhalant siphon. Sand particles tend to adhere to the periostracum of small individuals.

PUNCTATE PANDORA
Pandora punctata

OTHER NAME: Dotted pandora.
DESCRIPTION: Shell exterior and interior are white. Valves are crescent-shaped and flat with a curved upper edge.
SIZE: To 1 3/4" (4.5 cm) long.
HABITAT: In sand and mud of subtidal waters 7' (2 m) to 165' (50 m) deep.
RANGE: Esperanza Inlet, Vancouver Island, BC, to Punta Pequeña, Baja California Sur, México.
NOTES: Although this species is not an intertidal one, its empty shells occasionally turn up on the shoreline. The curious shape makes these shells easy to identify.

OCTOPODS AND SQUIDS
Class Cephalopoda

The mollusks in this group have several remarkable characteristics. A total of 8 or more arms are positioned around the mouth, and 2 gills, 2 kidneys and 3 hearts are also present. Some species release a dark fluid to aid in defense against predators.

TWO-SPOTTED OCTOPUS
Octopus bimaculoides

OTHER NAMES: Two spot octopus, California two spot octopus, mud flat octopus.

DESCRIPTION: Color varies from gray to brown, reddish or olive mottled with black, with **two bluish black eye spots** just below the real eyes. Arm length is 2 1/2 to 3 times the mantle length.

SIZE: Mantle length (from the end of the mantle to a point between the eyes) to 8" (20 cm). The mantle is a hood that contains the internal organs.

HABITAT: In holes or crevices of the mid-intertidal zone to water 65' (20 m) deep.

RANGE: San Simeon, San Luis Obispo County, California, to Ensenada, Baja California, México.

NOTES: The two-spotted octopus uses its radula (raspy tongue) to drill a hole in a variety of mollusks, including the black turban (see p. 81), purple olive (p. 107), Pacific littleneck (p. 150) and Pacific blue mussel (p. 126). Various small fishes are also captured for food, but some larger fishes are enemies. Another small species, the red octopus (*Octopus rubescens*), occurs from Alaska to Baja California. It is reddish in color but lacks the distinctive bluish eye spots of the two-spotted octopus.

LAMPSHELLS
Phylum Brachiopoda

Lampshells look somewhat like clams and were once classified as mollusks, but are now known to be a separate phylum. All lampshells are marine filter feeders that collect diatoms and detritus from the water with cilia (fine hair-like protrusions). The animal is protected by 2 shells and attached to a substrate with a stalk. Fossil records indicate that there were once more than 30,000 species, but today only 325 species survive.

COMMON LAMPSHELL
Terebratalia transversa

OTHER NAMES: Common Pacific brachiopod, lampshell.
DESCRIPTION: Color varies from gray to pinkish. Short stalk supports the hinged shells, which may or may not be ribbed.
SIZE: To 1" (2.5 cm) wide.
HABITAT: On rocks, low intertidal zone to subtidal waters 6,000' (1,829 m) deep.
RANGE: Alaska to Baja California, México.
NOTES: This brachiopod is thought to resemble an Aladdin's lamp, hence its name. Its chief enemies are crabs, which chip away at the shell to reach a tender meal inside. Fortunately for the common lampshell, it is usually attached to a rock in a covered or hidden location. A full 10 years is required for this brachiopod to reach its maximum size.

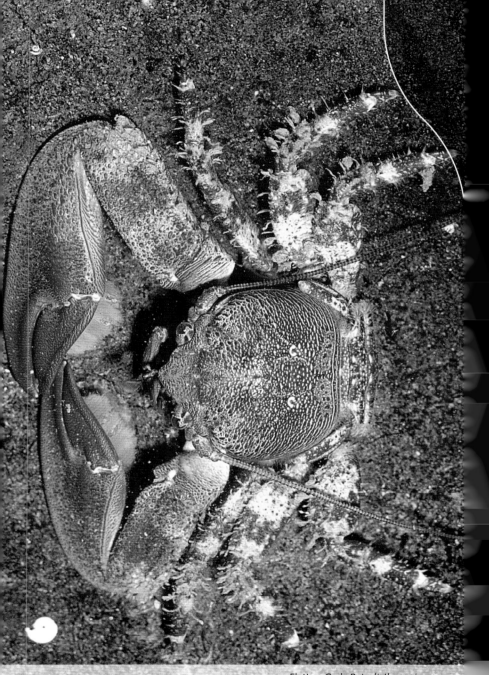

Flattop Crab *Petrolisthes eriomerus*.

ARTHROPODS
Phylum Arthropoda

ARTHROPODS

Arthropods are the most widespread group of creatures in the animal kingdom, as well as having the greatest number of species. Marine arthropods include barnacles, isopods, amphipods, shrimps and crabs. All members of this large group, phylum Arthropoda, have "jointed limbs" and an exoskeleton, or a skeleton that covers the body like armor.

BARNACLES
Class Cirripedia

Barnacles have modified legs (cirri) that sweep through the water like a net to collect tiny planktonic food. They reach sexual maturity at approximately 80 days and their reproduction is unusual: males may become females and vice versa at any time. To reproduce, the male must locate a female close enough for his penis to reach, as barnacles are unable to move from their substrate. The penis, however, can extend to 20 times the length of the barnacle's body.

LITTLE BROWN BARNACLE
Chthamalus dalli

OTHER NAMES: Small acorn barnacle, buckshot barnacle.
DESCRIPTION: Grayish brown shell. Relatively large central opening with a "+" pattern where cover plates meet.
SIZE: To 1/4" (6 mm) in diameter, 1/8" (4 mm) high.
HABITAT: Normally attached to rock in the **high intertidal zone**.
RANGE: Alaska to San Diego, California.
NOTES: This barnacle is distinct from others in that it does not crowd into spaces in a way that produces elongated individuals. But under ideal conditions, it can reach populations of 60,300 per square yard (72,000/m^2). This species is known to occur higher on intertidal rocks than any other barnacles, so it must tolerate very long periods of time out of water in the hot sun. Some barnacles can lose up to 40% of the water in their bodies in less than 9 hours. Another similar species, the brown buckshot barnacle *Chthamalus fissus*, is present from San Francisco to Baja California, México. The 2 species are very similar and can only be positively identified by dissection.

Red Thatched Barnacle

Tetraclita rubescens

OTHER NAMES: Volcano barnacle, thatched barnacle; formerly *Tetraclita squamosa*.
DESCRIPTION: Adults are **brick red** and volcano shaped; exterior is ridged or **thatched** in appearance. The opening is small. The young of this species are white.
SIZE: To 2" (5.1 cm) in diameter, 2" (5.1 cm) high.
HABITAT: On rock at exposed sites in the mid- to low intertidal zone and occasionally subtidally.
RANGE: Tomales Bay, Sonoma County, California, to San Lucas, Baja California, México.
NOTES: The red thatched barnacle prefers shady sites such as those found on overhangs or under ledges. Adults produce up to 3 broods during the summer after they reach 2 years of age. Each brood yields between 1,000 and 50,000 larvae. Adults may live as long as 15 years.

Acorn Barnacle

Balanus glandula

OTHER NAMES: Common acorn barnacle, *Balanus glandulus*.
DESCRIPTION: White to gray, cone-shaped. Younger living specimens have a visible dark black lining of cover plates; this is not easily seen in older individuals.
SIZE: To 3/4" (1.8 cm) in diameter, 3/8" (1 cm) high.
HABITAT: On rocks in **high to mid-intertidal zone**, and on various hard-shelled animals in both exposed and protected sites.
RANGE: Aleutian Islands, Alaska, to southern California.
NOTES: The acorn barnacle can grow into elongated columns under crowded conditions. This barnacle produces between 2 and 6 broods per year, during the cooler months. Individuals have been known to live to 15 years. A similar species, the little brown barnacle (p. 162), is smaller with brown coloration.

ARTHROPODS

RED-STRIPED ACORN BARNACLE
Balanus pacificus

DESCRIPTION: Outer walls are white with vertical **pink to purple stripes**. Shell is **smooth** or weakly ribbed.
SIZE: To 1 3/8" (3.5 cm) in diameter.
HABITAT: On hard surfaces from low intertidal zone to water 240' (73 m) deep.
RANGE: Monterey Bay, California, to Chile.
NOTES: The red-striped acorn barnacle shows a distinct preference to attach to other forms of marine life, including mollusk shells, exoskeletons of crustaceans and the eccentric sand dollar (see p. 204). This species, like all *Balanus*, has 6 plates in the outer shell.

WHITE-RIBBED RED BARNACLE
Megabalanus californicus

OTHER NAME: Red-striped acorn barnacle, red and white barnacle, pink barnacle; formerly *Balanus tintinnabulum*.
DESCRIPTION: Outer plates **pink to red with 12–15 white vertical ribs** that converge to a point.
SIZE: To 2 3/8" (6 cm) in diameter, 2" (5.1 cm) high.
HABITAT: From the low intertidal zone to water 30' (9 m) deep.
RANGE: Humboldt Bay, California, to Guaymas, México.
NOTES: This vividly colored barnacle occurs on rocks, kelp, mussels, crabs and other hard-shelled animals, and is sometimes found clustered on pilings and buoys. In crowded conditions it becomes very elongated. A similar species, the little striped barnacle *Balanus amphitrite*, is an introduced species. It is smaller, to 3/4" (1.9 cm) in diameter, has much less red striping and is usually found in quiet bays and estuaries.

GOOSE BARNACLE
Pollicipes polymerus

OTHER NAMES: Leaf barnacle, gooseneck barnacle; formerly *Mitella polymerus*.
DESCRIPTION: Cream-colored plates; generally dark brown body. Upper portion of body is supported by a peduncle (flexible stalk), which may grow to 6" (15 cm) in length. Individuals growing in the center of a group are the longest.
SIZE: To 4" (10 cm) long.
HABITAT: On exposed coast in areas subjected to strong wave action, mid-intertidal zone and lower.
RANGE: Sitka, Alaska, to Baja California, México.
NOTES: These barnacles often live in close association with the California mussel (see p. 125). Their resilient stalks are tough enough to withstand the forces tossing them in the surf, and their presence indicates you are in an area subject to harsh ocean waves. (Be ever watchful for unexpected waves when you are in such areas.) The goose barnacle is edible and has been exported from North America to Europe as a delicacy. Gulls, too, find this a tasty species and can consume it in large numbers.

PELAGIC GOOSE BARNACLE
Lepas anatifera

OTHER NAME: Common goose barnacle.
DESCRIPTION: Gray to bluish gray plates, orange-brown to purplish brown body with a brilliant scarlet-orange edge opening. An elongated stalk supports the flat, wedge-shaped body.
SIZE: To 6" (15 cm) long, $2^{3/4}$" (7 cm) wide.
HABITAT: Normally on driftwood, floating in the open ocean. Small individuals are sometimes found attached to seaweed, stranded on the beach.
RANGE: The oceans of the world.
NOTES: This gregarious barnacle is a creature of the high seas. The young are attracted to floating objects, which become home to hundreds or thousands of individuals. Once the "colonies" have been afloat for some time, they mature and produce their young. To observe this species on shore, look for a storm-stranded float, bottle or log on which the barnacles have settled.

ARTHROPODS

SHRIMPS, CRABS AND ALLIES
Class Malacostraca

Members of the shrimp and crab clan have several similar characteristics, including an exterior skeleton, jointed legs, two pairs of antennae and many body segments.

A hard exterior skeleton is one of the characteristics of crabs, so as they grow they must shed their shells periodically. The new shells are soft and remain so for a few days while the crabs grow rapidly, hence the term "soft-shelled crab." Crabs are most vulnerable during this transition period.

Some shrimps are hermaphrodites—male when they are small, changing to female when they become adults. Females carry their eggs and protect them until they hatch. Shrimp, like crabs, are the focus of a large commercial fishery.

Isopods and Amphipods

SCAVENGING ISOPOD
Cirolana harfordi

OTHER NAMES: Harford's greedy isopod, dark-backed isopod, swimming isopod.
DESCRIPTION: Color varies from gray to brown or blackish. Body is oval; distinctive triangular tailpiece has 2 flattened appendages.
SIZE: To 3/4" (2 cm) long.
HABITAT: Under rocks and seaweed and in mussel beds, from the high to low intertidal zone.
RANGE: BC to Baja California, México.
NOTES: This isopod, a common inhabitant of sandy and rocky shores, is a good swimmer and an important scavenger in its work cleaning up animal debris. Its importance is especially apparent when it reaches densities of 10,500 per square yard (12,600 per m^2). In fact, ichthyologists (fish biologists) sometimes use these animals to clean fish skeletons for study. The scavenging isopod is preyed on by several fishes.

Vosnesensky's Isopod
Idotea wosnesenskii

OTHER NAMES: Olive green isopod, green isopod, rockweed isopod, kelp isopod.
DESCRIPTION: Brown, green or red with a broad, flat body and a rounded tip on the abdomen.
SIZE: To 1 3/8" (3.5 cm) long.
HABITAT: On various seaweeds or under rocks, high intertidal zone to water 53' (16 m) deep.
RANGE: Alaska to San Diego, California.
NOTES: This isopod is very well adapted to living among seaweed and rocks. It is a master of disguise, usually cryptically colored and very difficult to find. Algae are the mainstay of its diet. This isopod was named in recognition of the work that the Russian zoologist Ilya Gavrilovich Vosnesensky conducted in Siberia, Alaska and California.

Western Sea Roach
Ligia occidentalis

OTHER NAME: Rock louse.
DESCRIPTION: Color is generally gray with orange-tipped legs. Species **resembles a cockroach**. Its **forked tail appendages are about 1/3 of its body length**.
SIZE: Body to 1" (2.5 cm) long.
HABITAT: On and under rocks from the splash zone to the high intertidal zone.
RANGE: Sonoma County, California, to Central America.
NOTES: Scurrying over rocks in the sunlight, this large isopod is hard to miss. It is most active at night and changes its color each morning to become darker during daylight. This isopod is a terrestrial species but it must live near a water source such as a tidepool. There it dips its rear end, where its gills are located, into water, in order to keep its breathing apparatus moist.

ARTHROPODS

CALIFORNIA BEACH HOPPER
Megalorchestia californiana

OTHER NAMES: California beach flea; formerly *Ocrhestoidea californiana*.
DESCRIPTION: Body is ivory-white, round and heavy. The **antennae are red in adults, orange in juveniles**, and longer than the body.
SIZE: To 1 1/8" (2.8 cm) long.

HABITAT: Along clean, fine sand beaches at the high tide mark and above.
RANGE: Vancouver Island, BC, to Laguna Beach, Orange County, California.
NOTES: This beach hopper is often found early in the morning, hiding among seaweed debris at the high tide mark once the tide recedes. At night, large numbers gather to feed on washed-up pieces of seaweed where the water laps onto the beach. Beach hoppers are sometimes called beach fleas, but they are not insects.

SKELETON SHRIMP
Caprella sp.

OTHER NAME: Phantom shrimp.
DESCRIPTION: Green or brown elongated body with 4 leg-like appendages and grasping claws.
SIZE: To 2" (5.1 cm) long.
HABITAT: On hydroids, eel-grass and seaweed near the low tide level and below in shallow water.
RANGE: BC to southern California.
NOTES: This species resembles a shrimp but it is not a shrimp, it is an amphipod. It is a fast-moving creature that feeds on tiny plants and animals. Females with obvious brood pouches can sometimes be found. The pouches are located on the third and fourth thoracic segments, as are the gills.

Shrimps

SMALLEYED SHRIMP

Heptacarpus carinatus

OTHER NAMES: Small-eyed coastal shrimp, smalleye coastal shrimp.
DESCRIPTION: Color varies widely from reddish brown to bright green, frequently with a white dorsal stripe. A very long and straight rostrum (nose-like blade) is present, as well as a prominent **beak-like projection on the third segment of the abdomen**. Its eyes are small.
SIZE: To $2^{3}/8$" (6 cm) long.
HABITAT: From tidepools in the low intertidal zone to water 89' (27 m) deep.
RANGE: Dixon Harbor, Alaska, to Point Loma, California.
NOTES: The smalleyed shrimp is often found among algae, surf-grass and eel-grass in tidepools along the exposed coast, with body colors closely matching that of the surrounding vegetation. It is often found in the company of several other shrimps. Females are larger than males, as in many shrimp species.

SITKA SHRIMP

Heptacarpus sitchensis

OTHER NAMES: Sitka coastal shrimp, common coastal shrimp; formerly *Heptacarpus pictus*.
DESCRIPTION: Normally green with red diagonal stripes and blue dots. Stout body includes an elongated rostrum (nose-like blade).
SIZE: To 1" (2.5 cm) long.
HABITAT: In gravel areas mixed with sand, and beneath rocks at low tide, mid-intertidal zone to 40' (12 m) deep.
RANGE: Alaska to Baja California, México.
NOTES: Sitka shrimp live in areas with little or no algae or eel-grass. Females have been seen carrying their eggs during May, June and September. Transparent individuals can be found on occasion.

ARTHROPODS

RED ROCK SHRIMP
Lysmata californica

OTHER NAMES: Striped rock shrimp; formerly *Hippolysmata californica*.
DESCRIPTION: Bold red-striped pattern on a translucent body, but this can change: color may shift to greenish, especially at night. First antennae are extremely long.
SIZE: To 2 3/4" (7 cm) long.
HABITAT: From tidepools in the low intertidal zone to water 200' (60 m) deep.
RANGE: Tomales Bay, Marin County, California, to Galapagos.
NOTES: The red rock shrimp feeds on a variety of foods as well as being a "cleaner" shrimp. It removes parasites and edible morsels from a variety of fish, the California spiny lobster (see p. 171) and even the fingernails of the occasional diver. From time to time, however, they are eaten by the animals they clean.

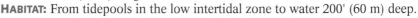

SMOOTH BAY SHRIMP
Crangon stylirostris

OTHER NAME: Smooth crangon.
DESCRIPTION: Coloration of this shrimp closely resembles the sand in which it burrows. Its carapace has no central spine.
SIZE: To 2 3/8" (6.1 cm) long.
HABITAT: In exposed sandy areas, intertidal zone to water 264' (80 m) deep.
RANGE: Alaska to Santa Cruz, California.
NOTES: This shrimp is capable of lightning-fast movements, which aids it greatly in avoiding predators. If you watch carefully, you will notice that this species prefers to bury itself in the sand whenever possible. It can hide itself almost completely by carefully pushing sand over its body with its long antennae, leaving only a tiny portion of its head uncovered. Very small clams and crustaceans form the bulk of this species' diet.

California Spiny Lobster

Panulirus interruptus

OTHER NAMES: California lobster, California rock lobster, red lobster.
DESCRIPTION: Body is reddish brown overall with **sharp red spines that cover most of the body**. There are no claws, and the antennae are very long.
SIZE: Normally to 16" (41 cm) long, although a few individuals have reached 36" (90 cm) long and 30 lbs (13.5 kg) in weight.

Exoskeleton of a California spiny lobster.

HABITAT: From the low intertidal zone to water 300' (91 m) deep.
RANGE: San Luis Obispo County, California, to Rosalina Bay, Baja California, México.
NOTES: This excellent-tasting species is fished both commercially and for sport in California. If you plan to harvest it, check current California Fish and Game regulations. Empty exoskeletons of the California spiny lobster are found on shore more often than living animals. The species hides by day and emerges at night to forage on a variety of both plants and animals. It is sometimes cleaned by the red rock shrimp (see p. 170), which is mutually beneficial since the shrimp obtains food while the lobster is thoroughly cleaned.

Bay Ghost Shrimp

Neotrypaea californiensis

OTHER NAMES: Ghost shrimp, sand shrimp, California ghost shrimp, *Callianassa californiensis*.
DESCRIPTION: White with **pink, yellow or orange** highlights. **One claw is enlarged**, especially in the male.
SIZE: To 4" (10 cm) long.
HABITAT: On sand or mud tide flats, mid- to low intertidal zone.
RANGE: Southeast Alaska to Baja California, México.
NOTES: The bay ghost shrimp lives to 10 years in its J-shaped burrow. A characteristic volcano-shaped mound surrounds each burrow entrance. These non-permanent burrows may be as deep as 30" (75 cm). At least 9 different tenant species are known to live in the burrows while the owner is present. This shrimp lays its eggs in the spring and they hatch from June through August.

ARTHROPODS

Crabs

MOONSNAIL HERMIT
Isocheles pilosus

OTHER NAMES: Formerly *Holopagurus pilosus*.
DESCRIPTION: Walking legs and claws are light yellow to orange with blotches of blue; antennae are light blue and hairy. Normally the left claw is slightly larger than the right.
SIZE: Carapace to $1^{1/4}$" (3 cm) long.
HABITAT: In sand or sandy mud from the low intertidal zone to 180' (55 m).
RANGE: Bodega Bay, California, to Baja California, México.
NOTES: The moonsnail hermit is a large species that buries itself into the sand so that only its eyes and antennae are exposed at the sand's surface. Although this hermit is often found in bays and estuaries, it can also be found in semi-exposed sites such as Pismo Beach. As its name suggests, it commonly inhabits empty moonsnail shells (see pp. 94, 95).

GRAINYHAND HERMIT
Pagurus granosimanus

OTHER NAME: Granular hermit crab.
DESCRIPTION: Dull green in color, with **blue or whitish spots or granules covering the body. Antennae are orange and unbanded**.
SIZE: Carapace to $3/4$" (1.9 cm) long.
HABITAT: In rocky or gravel areas and tidepools, low intertidal zone to water 118' (36 m) deep.
RANGE: Alaska to Baja California, México.
NOTES: This common species recycles larger, empty shells such as the black turban (see p. 81) and frilled dogwinkle (p. 99). The antics of this hermit crab, the clown of the seashore, can often be observed in tidepools. It reacts to the slightest movement by withdrawing into its shell, then falls from where it was and rolls to the bottom of its home, at which time it comes back out to do it all over again. This species is found lower intertidally than the hairy hermit crab (opposite).

Hairy Hermit

Pagurus hirsutiusculus

Description: Noticeable **narrow white band on lower portion of each walking leg**, sometimes with a blue spot on upper portion of the same segment. **Brownish antennae bear distinct bands**. Much of the **crab may be covered with hair**.
Size: To 3/4" (1.9 cm) long.
Habitat: In tidepools with sand or rock, in protected rocky areas, mid-intertidal zone to water as deep as 363' (110 m).
Range: Pribilof Islands, Alaska, to Monterey, California.
Notes: The hairy hermit chooses a variety of empty shells, including striped dogwinkle (see p. 100), purple olive (p. 107) and occasionally the turbans (*Tegula* spp., pp. 81–84). This hermit often abandons its shell altogether after it has been picked up. This provides an excellent opportunity to see its entire body, normally hidden inside the shell.

Blueband Hermit

Pagurus samuelis

Other name: Blue-handed hermit crab.
Description: Dull green, **bright blue bands circle each of the walking legs** near the tips. **Antennae are bright red and unbanded**.
Size: Carapace to 3/4" (1.9 cm) long.
Habitat: In somewhat exposed locations near rocks, often in tidepools, high intertidal zone.
Range: Vancouver Island, BC, to Punta Eugenia, Baja California, México.
Notes: The blueband hermit eats both plant and animal material. Most feeding activity occurs during the darkness of night. This hermit crab often occupies the abandoned shells of the black turban (see p. 81) and striped dogwinkle (p. 100). Shells constantly change ownership in the world of hermit crabs. As the crabs grow, they are ever watchful for new shells, even those currently being used by neighboring hermits. The young of the grainyhand hermit (see p. 172) are similar to this species' young, but their antennae are banded.

ARTHROPODS

UMBRELLA CRAB
Cryptolithodes sitchensis

OTHER NAMES: Turtle crab, butterfly crab.
DESCRIPTION: Color can include bright red, orange, gray or white. The flattened and flared carapace conceals the crab's legs when viewed from above. Shell also includes a distinctive rostrum (nose-like blade).
SIZE: Carapace to 3 1/2" (9 cm) wide.
HABITAT: On bedrock and in tidepools of the outer coast in protected to semi-exposed sites, low intertidal zone to water 56' (17 m) deep.
RANGE: Torch Bay, Alaska, to Point Loma, San Diego County, California.

NOTES: This slow-moving crab is a great treat to find at the lowest of tides. It remains motionless, and its many color phases help it blend in well with its surroundings, so it often goes unnoticed. The umbrella crab feeds on a wide variety of organisms, including calcareous red algae. Little else is known about the biology of this species.

FURRY CRAB
Hapalogaster cavicauda

OTHER NAMES: Fuzzy crab, hairy lithodid crab.
DESCRIPTION: Soft, golden brown to brown hair covers the upper body. A soft, sac-like abdomen is folded under the flattened body.
SIZE: Carapace to 3/4" (2 cm) wide.
HABITAT: Under rocks from the low intertidal zone to water 50' (15 m) deep.
RANGE: Cape Mendocino, Humboldt County, California, to Isla San Gerónimo, Baja California, México.
NOTES: The furry crab is both a filter feeder and an omnivore, which feeds actively on a variety of plants and animals. It is a curious crab that often remains motionless when the rock it is clinging to is turned over.

Granular Claw Crab

Dedignathus inermis

OTHER NAMES: Soft-bellied crab, papillose crab.
DESCRIPTION: Carapace is somewhat pear-shaped. The single large claw is covered in purple granular bumps.
SIZE: Carapace to 1" (2.5 cm) wide.
HABITAT: In rock crevices and occasionally in mussel beds, mid-intertidal zone to water 50' (15 m) deep.
RANGE: Alaska to Pacific Grove, Monterey County, California.
NOTES: This crab is well protected within crevices, its large claw often the only portion that is visible. A pair of these crabs often take up residence together in the same cavity. The crab uses its large, clumsy but powerful claw to crush small mussels. The smaller left claw is used effectively for grabbing, scraping and other feeding motions. The black oystercatcher *Haematopus bachmani* is known to prey upon this crab when it is found in mussel beds.

Flattop Crab

Petrolisthes eriomerus

OTHER NAMES: Porcelain crab, flat-topped crab.
DESCRIPTION: Reddish brown to blue-gray with a flat, nearly circular shell and long whip-like antennae. **Blue mouthparts** and **spot at the base of the movable part of the flat claws**. Length of carpus (wrist segment) of claw is about 2 times its width.
SIZE: Carapace to 3/4" (1.9 cm) wide.
HABITAT: Under rocks and among mussels in mussel beds, especially in areas with swift currents, low intertidal zone to water 284' (86 m) deep.
RANGE: Alaska to La Jolla, San Diego County, California.
NOTES: Flattop crabs are known to live together in groups of males, females and young. One dominant male does all or most of the breeding. The claws of this crab bear clusters of fine hairs, used to gather food from rock surfaces at night. Females often have 2 broods of young per year, each consisting of 10 to 1,580 eggs. This crab, like all porcelain crabs, has only 4 visible pairs of walking legs.

ARTHROPODS

FLAT PORCELAIN CRAB
Petrolisthes cinctipes

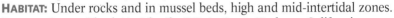

OTHER NAME: Smooth porcelain crab.
DESCRIPTION: Color varies from brown to blue. **Red mouthparts** and a **red spot at the base of the movable part of each claw**. Length of carpus (wrist segment) of claw is about $1^{1}/_{2}$ times as long as the width.
SIZE: Carapace to $^{7}/_{8}$" (2.4 cm) wide.
HABITAT: Under rocks and in mussel beds, high and mid-intertidal zones.
RANGE: Queen Charlotte Islands, BC, to Santa Barbara, California.
NOTES: This crab is sometimes found in high numbers in beds of the California mussel (see p. 125). In some ideal locations, their numbers have been calculated to reach 1,029 individuals per square yard (860 per m^2). This crab, like all porcelain crabs, sheds its brittle claws or legs easily when it feels threatened. The missing appendages grow back after several molts.

CABRILLO'S PORCELAIN CRAB
Petrolisthes cabrilloi

OTHER NAME: Porcelain crab.
DESCRIPTION: A light brown crab with **red mouthparts and a red spot at the base of the movable finger on each claw. Hair covers the carpus (wrist segment) and walking legs**.
SIZE: Carapace to $^{5}/_{8}$" (1.6 cm) wide.
HABITAT: Under rocks and among mussel beds from the high to low intertidal zone.
RANGE: Morro Bay, San Luis Obispo County, California, to Bahía Magdalena, Baja California, México.
NOTES: Cabrillo's porcelain crab is common in southern California where it replaces the flat porcelain crab (see above) of the north. Females with eggs have been observed in February, April, May, October and November, but little is known about the natural history of this species.

Chocolate Porcelain Crab

Petrolisthes manimaculis

OTHER NAMES: Porcelain crab.
DESCRIPTION: Color brown with a row of **tiny blue spots on each claw**. Mouthparts are bright blue and there is a **red spot at the base of the movable finger** of the claw. Claws are long with the **carpus** (wrist segment) of claw about 2 1/2 to 3 times longer than it is wide.
SIZE: Carapace to 3/4" (2 cm) wide.
HABITAT: Under rocks in the low intertidal zone.
RANGE: Bodega Bay, Sonoma County, California, to Punta Eugenia, Baja California, México.
NOTES: The chocolate porcelain crab is known to drop its legs very easily when handled. These lost limbs are later regenerated. This may be an advantage since this species lives under rocks, which may shift onto a limb and crush it.

Thick-Clawed Porcelain Crab

Pachycheles rudis

OTHER NAME: Thickclaw porcelain crab.
DESCRIPTION: Dull brown with a rounded body and granule-covered claws.
SIZE: Carapace to 3/4" (2 cm) wide.
HABITAT: Normally in pairs under rocks, in the empty burrows of rock-dwelling clams and similar sheltered locations, low intertidal zone to water 95' (29 m) deep.
RANGE: Kodiak, Alaska, to Baja California, México.
NOTES: This crab uses its specialized mouthparts to filter out plankton and other microscopic food from the water. Unlike many types of crabs, male and female thick-clawed porcelain crabs grow to the same size.

ARTHROPODS

PACIFIC MOLE CRAB
Emerita analoga

OTHER NAME: Pacific sand crab.
DESCRIPTION: Tan to gray with an egg-shaped body that lacks claws.
SIZE: To 1 3/8" (3.5 cm) long.
HABITAT: This surf-loving crab moves up and down the beach with the tides.
RANGE: Oregon to Chile, with a few records as far north as Kodiak, Alaska.

NOTES: This crab burrows into the sand backwards so that only its head and antennae are exposed at the surface. It uses its long, feathery antennae to catch plankton and detritus, its prime food. If disturbed, these crabs disappear into the sand in mere seconds—truly amazing to watch. Temporary populations have been found north of Oregon. These begin when larvae reach new sites via ocean currents, then individuals successfully grow to maturity. These crabs cannot reproduce, however, and the colony eventually disappears.

GRACEFUL DECORATOR CRAB
Oregonia gracilis

OTHER NAME: Decorator crab.
DESCRIPTION: Carapace is drab and often too well camouflaged to be visible. Triangular shell is pointed at the front, wide and rounded at the back. A pair of long horns elongate the "snout area." Legs are noticeably elongated.
SIZE: Carapace to 1 1/2" (3.9 cm) wide.
HABITAT: Among seaweed, intertidal zone to water 1,439' (436 m) deep.
RANGE: Bering Sea to Monterey Bay, California.
NOTES: This species is known for its elaborate camouflage, which is made of seaweeds, hydroids, sponges, bryozoans and virtually anything available. These it carefully fastens on its upper shell or carapace and legs with small, curved setae (velcro-like hooks).

Graceful Kelp Crab
Pugettia gracilis

OTHER NAMES: Kelp crab, graceful rock crab, slender crab.
DESCRIPTION: Highly variable color, ranging from white to bright red, often ornamented with small amounts of seaweed, sponges or bryozoans. Fingers of claws are blue or gray, tipped with orange.
SIZE: Carapace to $1^{1/2}$" (4 cm) wide.
HABITAT: Among rocks and algae, from low intertidal zone to water 460' (140 m) deep. Young individuals are often found among eel-grass.
RANGE: Aleutian Islands, Alaska, to Monterey, California.
NOTES: This crab's most noticeable feature is the decorations attached to its shell. As with all crabs, the eggs are kept on the female until the young are ready to hatch. They are attached in such a way that oxygen can reach each egg while water circulates.

Shield-Backed Kelp Crab
Pugettia producta

OTHER NAMES: Kelp crab, northern kelp crab.
DESCRIPTION: Generally olive in color, with varying amounts of red or yellow in adults. Distinctive smooth, **shield-shaped carapace**.
SIZE: Carapace to $3^{5/8}$" (9.3 cm) wide.
HABITAT: Usually in or on seaweed, especially bull kelp (see p. 233), intertidal zone to water 240' (73 m) deep.
RANGE: Alaska to Baja California, México.
NOTES: This crab feeds primarily on kelp or large brown seaweed, but will feed on a variety of organisms if kelp is not available. The smooth carapace closely matches the algae it is often found upon. In the fall, adults migrate to kelp in deeper water. Here they congregate to feed and mate until December, when they return to shallow waters. Occasionally barnacles grow on the backs of adults, but this species does not camouflage itself as other related species do. Be sure to check pilings while at the dock. Large adults can often be found there.

ARTHROPODS

CRYPTIC KELP CRAB

Pugettia richii

DESCRIPTION: Color varies widely from dark brown to red. Teeth on edge of carapace project laterally, each with a more pronounced curve than the last. Fingers of claws are white.
SIZE: Carapace to 1 3/4" (4.2 cm) wide.
HABITAT: Under rocks from the low intertidal zone to water 325' (100 m) deep.
RANGE: Prince of Wales Island, Alaska, to Isla San Gerónimo, Baja California, México.
NOTES: Like other kelp crabs, this species adorns itself with bits of seaweed, bryozoans and hydroids. The cryptic kelp crab, however, tends to decorate only its rostrum (nose-like blade) area. They may also become abundant in kelp holdfasts.

GLOBOSE KELP CRAB

Taliepus nuttallii

OTHER NAMES: Southern kelp crab; formerly *Epialtus nuttallii*.
DESCRIPTION: Color varies from purple to dark red-brown. Carapace is noticeably convex with a prominent rostrum (nose-like blade) on the shell.
SIZE: Carapace of males to 3 3/4" (9.2 cm) wide; females are about half that size.
HABITAT: From the low intertidal zone to subtidal waters 300' (93 m) deep.
RANGE: Santa Barbara, California, to Bahía Magdalena, Baja California Sur, México.
NOTES: The globose kelp crab prefers to feed on large brown algae such as giant perennial kelp (see p. 234) and feather boa kelp (see p. 231). This crab may be found in surge channels or large tidepools at low tide. Its claws are often large and can deliver a nasty nip.

Dwarf Teardrop Crab
Pelia tumida

OTHER NAME: Dwarf crab.
DESCRIPTION: Color is generally yellow-brown due to attached sponges. Carapace is pear-shaped. The male's claws are a bright red.
SIZE: Carapace to 1/2" (1.5 cm) wide.
HABITAT: Under rocks and in kelp holdfasts from the low intertidal zone to water 330' (100 m) deep.
RANGE: Monterey, California, to Bahía de Petatlán, México and the Gulf of California.
NOTES: This distinctive arthropod is a slow mover that holds on tightly to its substrate when disturbed. The dwarf teardrop crab is pubescent (covered in fine, tiny hair); sponges often attach to the hairs. Algae sometimes attach to the carapace as well.

Helmet Crab
Telmessus cheiragonus

OTHER NAME: Horse crab.
DESCRIPTION: Yellowish brown, sometimes with red or orange added as well. Characteristic tiny hairs cover all surfaces.
SIZE: Carapace to 3 1/4" (8.3 cm) long, 4" (9.7 cm) wide.
HABITAT: Intertidal zone to 360' (110 m) deep.
RANGE: Norton Sound, Alaska, to Monterey, California.
NOTES: During the early spring, this crab is often observed intertidally during the breeding season, typically in areas with an abundance of seaweed in which the crab can hide. It is known to feed on eel-grass, algae, snails, bivalves and worms.

ARTHROPODS

RED ROCK CRAB
Cancer productus

OTHER NAMES: Red crab, red cancer crab.
DESCRIPTION: Brick red in color. Shell has smooth, saw-like outline on front, and 5 equal rounded "scallops" between the eyes. **Tips of claws are black.**
SIZE: Carapace **to 7" (18 cm) wide.**
HABITAT: Lower intertidal zone to water 260' (79 m) deep.
RANGE: Alaska to Baja California, México.
NOTES: The red rock crab, like all of the *Cancer* clan, is a carnivore. Its heavy claws are strong enough to crack open the shells of barnacles and snails. It is an opportunist, feeding also on small living crabs and dead fish. The young of this native species are often found in the intertidal zone. Their coloration ranges from stripes of various colors to near-white, but the shape of this crab does not change. The **outline of its carapace resembles the letter *D*.** The red rock crab is popular in the sport fishery.

PYGMY ROCK CRAB
Cancer oregonensis

OTHER NAMES: Hairy cancer crab, Oregon cancer crab, Oregon rock crab.
DESCRIPTION: Dark red with **round carapace** and hairy legs. **Claws are tipped with black.**
SIZE: Carapace **to 1 3/4" (4.7 cm) wide.**
HABITAT: Under rocks, low intertidal zone to 1,435' (435 m) deep.
RANGE: Pribilof Islands, Alaska, to Los Angeles, California.
NOTES: Mating takes place in the spring after the females molt. Courtship behavior includes males carrying females prior to their molting and continuing to carry them, after mating, until their new shells have hardened. Females store the sperm until late fall or winter. Females spawn in February, each carrying as many as 33,000 eggs. Barnacles are the primary food of this crab.

Spot-Bellied Rock Crab

Cancer antennarius

OTHER NAMES: Pacific rock crab, brown rock crab, common rock crab.
DESCRIPTION: Carapace is orange-brown above with **red spots on the underside**. Large **claws with black tips**. Shell is widest at the 8th or 9th tooth.
SIZE: Carapace to 7" (17.8 cm) wide.
HABITAT: On rocky shores from the low intertidal zone to water 300' (91 m) deep.
RANGE: Queen Charlotte Sound, BC, to Cabo San Lucas, México.
NOTES: The spot-bellied rock crab feeds on a wide variety of animals, including other crabs, snails and carrion. It has been observed using its large claws to chip away at hermit crabs' shells and reaching inside. These large claws can also deliver a painful nip to an unwary beachcomber! During the winter months, females are often encountered carrying their eggs on their underside.

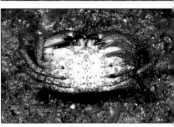

Dungeness Crab

Cancer magister

OTHER NAMES: Edible cancer crab, commercial crab, Pacific crab.
DESCRIPTION: Red-brown to **purple shell** with spine-tipped edge on front half. Shell is widest at the 10th tooth. **Claws have white tips**.
SIZE: Carapace to 9" (22.5 cm) wide.
HABITAT: On sandy bottoms, low intertidal zone to water 759' (230 m) deep.
RANGE: Alaska to Santa Barbara, California.
NOTES: The Dungeness crab is harvested both commercially and recreationally. Females lay up to 2.5 million eggs and this species is known to live at least 6 years. It is an active carnivore that feeds on at least 40 different species, including shrimp, small clams, oysters, worms and fish. Dungeness crabs spend a great deal of time buried under the sand. In the spring, they can often be found in the low intertidal zone buried in sand in tidepools. This is where they hide while waiting for their new shells to harden.

Underside of male.

Underside of female.

ARTHROPODS

GREEN CRAB
Carcinus maenas

OTHER NAMES: European green crab, shore crab, European shore crab.
DESCRIPTION: Color is greenish, mottled with black above and yellowish below. **5 large teeth are on outer edge of carapace and 3 teeth between eyes**. Last pair of legs are relatively flattened.
SIZE: Carapace to $3^{3/4}$" (9.2 cm).
HABITAT: On rocks, mud flats of sheltered bays and estuaries, and tidepools from the intertidal zone to water 20' (6 m) deep.
RANGE: Southern Vancouver Island, BC, to Elkhorn Slough, California.
NOTES: The green crab was accidentally introduced to San Francisco Bay in 1989 and since then it has rapidly expanded. It is an adaptable invader, native to Europe but has been introduced to many areas around the globe including Australia, South Africa and both coasts of the USA. Often described as a voracious omnivore, the green crab feeds on a wide variety of plants and animals, including clams, mussels, other crabs, small fish and almost anything else.

BLACK-CLAWED CRAB
Lophopanopeus bellus

DESCRIPTION: Carapace varies in color from orange to purple or white. A total of 3 spines project forward on the sides of the carapace, **claw fingers are usually dark**.
SIZE: Carapace to $1^{1/2}$" (4 cm) wide.
HABITAT: In sand or gravel, low intertidal zone to water 264' (80 m).
RANGE: Alaska to San Diego, California.
NOTES: The black-clawed crab eats a variety of food, including various algae, mussels and barnacles. Females are known to have 2 broods of young each year, each consisting of 6,000 to 36,000 eggs.

Retiring Southerner
Pilumnus spinohirsutus

OTHER NAMES: Hairy crab, hairy under-rock crab.
DESCRIPTION: Light brown hair covers most of the body. The claws have dark fingers and there are 5 curved spines on the edge of the carapace.
SIZE: Carapace to 1 3/8" (3.5 cm) wide.
HABITAT: Under rocks and buried in sand from the low intertidal zone to water 80' (25 m) deep.
RANGE: San Pedro, Los Angeles County, California, to Baja California, México.
NOTES: The behavior of captive specimens suggests that the retiring southerner is a carnivore. It is an inhabitant of rocky shores, often found where scaly tube snails (see p. 86) are especially abundant. This is a slow-moving crab, which uses its coloration to hide from predators. It occurs in warm–temperate waters, and as such is becoming less common because of the dredging of harbors.

Striped Shore Crab
Pachygrapsus crassipes

DESCRIPTION: Most often green, but may be red or purple with **black stripes across front**.
SIZE: Carapace to 1 3/4" (4.7 cm) wide.
HABITAT: In tidepools, high to mid-intertidal zone.
RANGE: Oregon to Gulf of California.
NOTES: This crab is often found underwater in tidepools. Well adapted to life outside the water, it comes out in large numbers to feed, especially among eelgrass. Its primary food is algae, which it picks up and brings to its mouth with alternating claws, seemingly shoveling it in with salad forks! There is also a report of this lively species capturing flies at low tide. Its main predators are gulls and raccoons. Striped shore crabs engage in interesting courtship behavior. The female releases a chemical (pheromone) to attract males. Then a sort of dance begins, in which a male turns over onto his back and the female walks over him. Approximately 50,000 eggs are eventually produced by the female, once or twice each year. This crab, a native of North America, is often said to have been introduced from Asia, but in fact it was introduced to Asia in the late 1800s.

ARTHROPODS

PURPLE SHORE CRAB

Hemigrapsus nudus

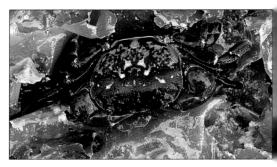

OTHER NAME: Purple rock crab.
DESCRIPTION: Color varies from purple to red-brown, with **red-purple spots usually present** on the claws. **No hair**, spines or other coverings on shell or legs (hence its name, *nudus*)—a useful feature for identification.
SIZE: Carapace to $2^{1/4}$" (5.6 cm) wide.
HABITAT: Under rocks and among seaweed, mid- to low intertidal zone. Often found out of the water on the shore.
RANGE: Alaska to Turtle Bay, México.
NOTES: The purple shore crab feeds mainly at night, consuming green algae such as sea lettuce (see p. 223). When this crab is discovered under a rock, it often walks sideways in an effort to escape and find a new hiding spot. Predators of adult crabs include the glaucous-winged gull *Larus glaucescens* and white-winged scoter *Melanitta fusca*.

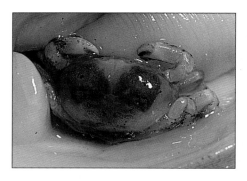

GAPER PEA CRAB

Pinnixa littoralis

OTHER NAME: Pea crab.
DESCRIPTION: White with dark markings. Dark markings are especially visible in the male.
SIZE: Carapace to 1" (2.5 cm) wide.
HABITAT: In the mantle cavity of several larger clams from the low intertidal zone to subtidal waters 300' (91 m) deep.
RANGE: Prince William Sound, Alaska, to Baja California, México.
NOTES: Gaper pea crabs live in pairs, inside the cavities of a clam in such a way that they do not harm the host. The female is much larger than the male. The young are sometimes found in the cavities of many species of clams as well as in larger limpets. Adults may be found in various larger clams, including Nuttall's cockle (see p. 137) and the Pacific gaper (p. 138).

MEXICAN FIDDLER
Uca crenulata

OTHER NAMES: California fiddler, fiddler crab.
DESCRIPTION: Carapace is tan to brown in color and rectangular. Eyestalks are elongated. Male has one enlarged pincer and a ridge on the inside of the hand. Pincers are equal size in females.
SIZE: Carapace to 3/4" (1.9 cm) wide.
HABITAT: In sandy mud flats of bays and estuaries in the high to mid-intertidal zone.
RANGE: Goleta Slough, California, to Isla de Ios Mangles, Baja California, México.
NOTES: The Mexican fiddler is the only fiddler found in California and is now considered rare in the USA. This fiddler makes a permanent burrow that may go as deep as 4' (1.2 m) into the sandy mud. The male uses its enlarged pincer in courtship rituals. The small pincer of the male and both pincers of the female are used for feeding on tiny organic particles found on the mud flats. The Mexican fiddler can be found at Upper Newport Bay.

A male at its burrow entrance.

ARACHNIDS
Class Arachnida

RED VELVET MITE
Neomolgus littoralis

OTHER NAME: Intertidal mite.
DESCRIPTION: Bright scarlet red. Short, dense hairs cover this mite, but cannot be seen without a microscope.
SIZE: To 1/8" (3 mm) long.
HABITAT: High intertidal zone.
RANGE: Alaska to Laguna Beach, Orange County, California.
NOTES: The red velvet mite is related to spiders and ticks, which have 8 legs rather than 6, as on insects. It is often seen scurrying about the high intertidal zone, in the heat of the day. This species has been observed using its snout-like mouthparts to feed on the fluids of dead flies. Additional studies are needed to determine the natural history of this brightly colored arachnid.

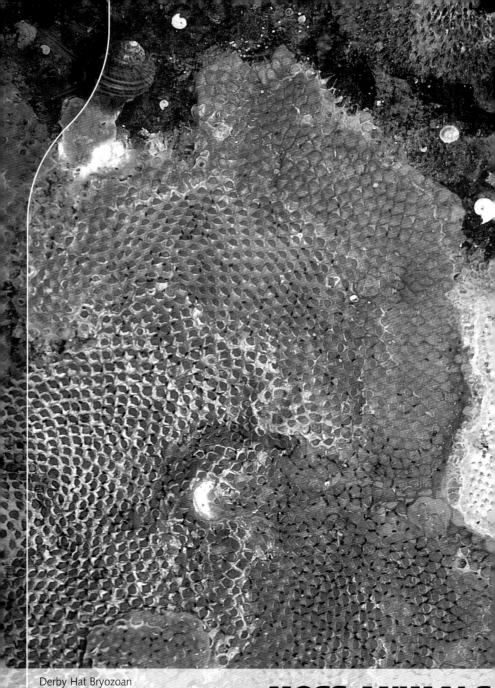

Derby Hat Bryozoan
Eurystomella bilabiata.

MOSS ANIMALS
Phylum Bryozoa

Moss animals or bryozoans are a group of nearly 2,000 different species that are often misidentified. They live hanging from marine algae, encrusting on rocks and shells, or growing upright from rock crevices. A bryozoan colony comprises thousands of individuals. The colony reproduces and grows by budding. Some species are rigid, others are flexible and sway in the water, still others are gelatinous. The bryozoan has a primitive nerve system but lacks a heart and vascular system.

BRANCHED-SPINE BRYOZOAN

Flustrellidra corniculata

OTHER NAME: *Flustrella cervicornis.*

DESCRIPTION: Tan to brown with numerous spines covering the colony, giving it a fuzzy appearance.

SIZE: Colony to 4" (10 cm) long.

HABITAT: Often found on various seaweeds, low intertidal zone to subtidal waters 248' (75 m) deep.

RANGE: Alaska to Point Buchon, San Luis Obispo County, California.

NOTES: The distinctive branched-spine bryozoan is a cold water species, also found in northern Europe. It looks somewhat like a fuzzy seaweed and has a soft, leather-like texture.

MOSS ANIMALS

KELP ENCRUSTING BRYOZOAN
Membranipora membranacea

OTHER NAMES: Encrusting bryozoan, lacy-crust bryozoan, kelp lace.
DESCRIPTION: White in color, forming thin, flat colonies with small, rectangular cell-like structures. Individuals tend to crowd together, radiating from the center.
SIZE: Colony to more than 3" (7.6 cm) in diameter.
HABITAT: On kelp fronds, floats and rocks.
RANGE: Alaska to Baja California, México, and all temperate regions of the world.
NOTES: The kelp encrusting byrozoan is found most often in the spring and fall. It commonly occurs on several species of kelp or brown algae, including bull kelp (see p. 233). This beautifully patterned species starts growing from the center and radiates outward from the oldest portion. The cryptic nudibranch (p. 113) sometimes lives on this bryozoan, but careful inspection is required to see it.

SEA-LICHEN BRYOZOAN
Dendrobeania lichenoides

OTHER NAME: Sea mat bryozoan.
DESCRIPTION: Tan or brown with numerous petal-like, irregularly shaped fronds.
SIZE: Fronds to 1" (2.5 cm) in diameter.
HABITAT: In shaded, protected, rocky areas, low intertidal zone to subtidal waters 300' (90 m) deep.
RANGE: Alaska to southern California.
NOTES: This species attaches to a wide variety of substrates including shells, rocks and worm tubes. Countless numbers of microscopic tentacles cover bryozoans. These tentacles are used to filter feed on phytoplankton, bacteria and detritus. The food is then moved to the central mouth of each individual by tiny hairs on the tentacles.

FLUTED BRYOZOAN

Hippodiplosia insculpta

DESCRIPTION: Yellow-brown to orange with double-fluted fronds.
SIZE: Colonies to 5" (12.5 cm) or more in diameter.
HABITAT: Attached to rocks, mid-intertidal to subtidal water 772' (234 m) deep.
RANGE: Gulf of Alaska to Gulf of California and Costa Rica.
NOTES: This bryozoan is hard to the touch. In northern colonies, where the water is much cooler, larger individuals (zooids) may be found.

DERBY HAT BRYOZOAN

Eurystomella bilabiata

OTHER NAME: Rosy bryozoan.
DESCRIPTION: Rose red to orange-red with a distinctive hat-shaped opening for each individual.
SIZE: Colony to 2" (5 cm) in diameter.
HABITAT: On stones or shells, low intertidal zone to subtidal waters 211' (64 m) deep.
RANGE: Alaska to Bahía de Tenacatita, México.
NOTES: This is an encrusting species in which many individuals (zooids) live side by side on a substrate. Each is enclosed in a case adjacent to its neighbors'. The colony expands when individuals bud off, producing new animals.

Western Spiny Brittle Star
Ophiothrix spiculata.

SPINY-SKINNED ANIMALS
Phylum Echinodermata

The echinoderms (spiny-skinned) are a large group of animals. All members of this group have calcareous plates covered with a soft layer of skin. The size of the plates varies from large and conspicuous, as in most species (e.g. sea urchins), to inconspicuous (e.g. sea cucumbers). Locally the echinoderms consist of sea stars, brittle stars, sea urchins, sand dollars and sea cucumbers.

SEA STARS
Class Asteroidea

Sea stars were once referred to as starfish, but sea star is a much better name as no individuals in this group swim, have scales or are edible. Sea stars feed on a wide variety of foods. Some of the more active species can actually capture live snails or other stars, while some slower species feed on various seaweeds. Movement is made possible by many small tube feet on the underside of each ray.

Animals in this class have truly remarkable powers of regeneration. Entire limbs can be regenerated and in some species, a whole sea star can be regenerated from a single ray with a portion of the central disc or body.

BAT STAR
Asterina miniata

OTHER NAMES: Broad-disc sea star, sea bat, webbed sea star; formerly *Patiria miniata*.
DESCRIPTION: Has been observed in nearly every color of the rainbow. Normally 5 webbed arms, but sometimes 4–9 arms.
SIZE: To 6" (15 cm) in diameter.
HABITAT: Normally on exposed sites with rock bottoms covered in surf-grass, algae, sponges and bryozoans, low intertidal zone to water 960' (290 m) deep.
RANGE: Sitka, Alaska, to Baja California, México.
NOTES: The bat star has been kept in captivity for more than 16 years and is thought to live up to 30 years. It is an omnivore, feeding on plants and animals both dead and alive. The bat star worm (see p. 44), a segmented worm, is often found living commensally on the underside of this sea star.

SPINY-SKINNED ANIMALS

LEATHER STAR
Dermasterias imbricata

OTHER NAME: Garlic star.
DESCRIPTION: Mottled red-brown to orange, with 5 rays. A slippery secretion covers the surface.
SIZE: To 10" (25 cm) in diameter.
HABITAT: On rocky shores, low intertidal zone to water 300' (91 m) deep.
RANGE: Prince William Sound, Alaska, to southern California.

Underwater, in a tidepool.

NOTES: Tidepoolers can easily recognize this common species. The leather star feels like wet leather and often smells like garlic or sulfur. Its diet includes the giant green anemone (see p. 30), proliferating anemone (p. 31), red-beaded anemone (p. 32), purple sea urchin (p. 203) and several other invertebrates, which it swallows whole and digests internally.

STRIPED SUN STAR
Solaster stimpsoni

OTHER NAMES: Sun star, Stimpson's sun star.
DESCRIPTION: Blue, pink, red or orange with a **blue or purple stripe down each slender arm**. Usually 10 arms, occasionally only 9.
SIZE: To 16" (40 cm) across.
HABITAT: In rocky areas, very low intertidal zone to subtidal waters 2,013' (610 m) deep.
RANGE: Bering Sea to Salt Point, Sonoma County, California.
NOTES: The striped sun star feeds on a variety of invertebrates, including sea cucumbers, lampshells and tunicates. This species is occasionally found in the intertidal zone.

Pacific Blood Star
Henricia leviuscula

OTHER NAMES: Blood star, Pacific henricia.
DESCRIPTION: Color varies widely from blood red to tan, yellow, orange and purple. Typically 5 slender arms, occasionally 4–6.
SIZE: To 8" (20 cm) in diameter.
HABITAT: On or under rocks covered with growth (sponges, etc.), in protected areas from low-tide mark to water 1,425 ft (435 m) deep.
RANGE: Aleutian Islands, Alaska, to Baja California, México.
NOTES: This common star feeds primarily upon sponges. Small females brood their bright red-orange eggs in darkness from January through March. Larger individuals do not brood their eggs at all, but release them directly into deeper water. Beachcombers are drawn to this star's vivid color, but if you take it home, the star often does not survive and its color fades when it dries. Rather than removing it from its habitat, try taking a photograph or making a sketch.

Six-Rayed Star
Leptasterias hexactis species complex

OTHER NAMES: Broad six-rayed sea star, six-armed sea star.
DESCRIPTION: Color varies from green to black, brown, orange, yellow or red, often with a pattern. Normally has 6 rays.
SIZE: To 4" (10 cm) in diameter.
HABITAT: On rocky shores, high intertidal to subtidal zone and below. Frequently found under rocks in intertidal areas.
RANGE: Vancouver Island, BC, to Santa Catalina Island, California.
NOTES: This small star is often found tightly attached to the underside of a rock. It feeds on sea cucumbers, barnacles, chitons, mussels, limpets and snails. Most feeding occurs during the summer months; little food is eaten over the winter months. The female broods her egg mass under her disc for 6–8 weeks before the eggs hatch. This star reaches maturity at approximately 2 years. Research suggests the six-rayed star may actually be several closely related species.

SPINY-SKINNED ANIMALS

GIANT PINK STAR
Pisaster brevispinus

OTHER NAME: Short-spined sea star.
DESCRIPTION: Characteristically pink to almost white. Short spines cover the dorsal side.
SIZE: To 26" (64 cm) in diameter.
HABITAT: Sand or mud bottom, low intertidal zone to 330' (100 m).
RANGE: Vancouver Island, BC, to Monterey Bay, California.
NOTES: This large sea star feeds on the Pacific geoduck (see p. 153) and eccentric sand dollar (p. 204),

among other species. The size of the individual sea star determines the size of its prey. This species is very similar to the purple star (p. 197), which is a smaller size and a purple or ocher color.

GIANT SPINED STAR
Pisaster giganteus

OTHER NAMES: Knobby sea star, knobby starfish, giant sea star.
DESCRIPTION: Color varies from brown to red or purple with a **blue ring surrounding each white-tipped spine**. Spines are blunt and scattered randomly.
SIZE: To 10" (25 cm) in diameter at intertidal sites and to 22" (56 cm) at subtidal locations.
HABITAT: On rocks and pilings in protected coastal sites from the very low intertidal zone to water 300' (90 m) deep.
RANGE: Vancouver Island, BC, to northern Baja California, México.
NOTES: The giant spined star feeds on a variety of marine life, including the California mussel (see p. 125), Pacific blue mussel (p. 126), sand-castle worm (p. 50) and various snails. It is known to spawn in March or April in the Monterey Bay area. This species is only found intertidally in the southern portion of its range.

PURPLE STAR
Pisaster ochraceus

OTHER NAMES: Ocher star, common sea star, Pacific sea star.
DESCRIPTION: Has **3 color phases: purple, brown and yellow**. Normally **5 stout arms**, but occasionally 4–7. A loose network of white calcareous plates stiffens the body. Dorsal blunt spines form a pentagon or reticulated pattern.
SIZE: To 14" (35 cm) in diameter.
HABITAT: Along **exposed** and protected rocky shorelines, mid- to lower intertidal zone to water 300' (90 m) deep.
RANGE: Alaska to Baja California, México.
NOTES: This is the most common sea star found in our intertidal zones. It feeds on mussels, abalone, chitons, barnacles and snails. Many of its prey species can detect this star when it is nearby and escape. But its favorite prey, including the California mussel (see p. 125) and the goose barnacle (p. 165), are attached and are unable to escape.

SUNFLOWER STAR
Pycnopodia helianthoides

OTHER NAME: Twenty-rayed sea star.
DESCRIPTION: Yellow, orange, brown, pink, red or purple. Typically 24 arms and a broad disc covered with a soft skin.
SIZE: Normally to 39" (1 m) in diameter; occasionally even larger.
HABITAT: On soft bottoms and rocky shores, low intertidal zone to 1,440' (437 m).
RANGE: Prince William Sound, Alaska, to southern California.
NOTES: A good-sized sunflower star has an estimated 15,000 tube feet on its body. It is the largest and fastest sea star found in California and is a voracious feeder, preying on many large clams and crustaceans. It has also been observed to feed on dead squid. It engulfs the indigestible squid pen, which is too large to pass normally from the star and which is extruded through the soft upper part of the star's body.

SPINY-SKINNED ANIMALS

BRITTLE STARS
Class Ophiuroidea

Most brittle stars shun light and spend the daylight hours hiding under rocks or in similar situations. Most species live in the tropics, but several species inhabit California's seashores. When stressed, these delicate creatures often shed their arms, then regenerate new ones in time.

DAISY BRITTLE STAR
Ophiopholis aculeata

OTHER NAMES: Serpent star, painted brittlestar.
DESCRIPTION: Occurs in a wide range of colors and patterns. **Central disc is a scallop shape with bulges between the arms.**
SIZE: Disc to 7/8" (2.2 cm) in diameter. **Length of arms to 3 1/2–4 times diameter of disc.**
HABITAT: Under stones, in algal holdfasts and rocky shores, low intertidal zone to water 5,465' (1,657 m) deep.
RANGE: Bering Sea to south of Santa Barbara, California.
NOTES: This common species, whose name comes from the flower-like shape of its disc, is more abundant in the northern part of its range. Like other brittle stars, it feeds by scraping minute organisms from rock with its specialized tube feet. The food then enters the stomach, which takes up most of the body cavity. Unlike the sea star, the brittle star cannot extrude its stomach to feed. Strangely enough there is no intestine or anus; food is absorbed along the alimentary canal and wastes go back out the mouth.

Dwarf Brittle Star

Ampipholis squamata

Other Names: Small brittle star, serpent star, holdfast brittle star, brooding brittle star, *Axiognathus squamatus*.
Description: Gray, tan or orange. A white spot is present near the base of each arm on the **round disc, which lacks bulges**.
Size: Disc to $3/16"$ (5 mm) in diameter. **Length of arms to 3–4 times diameter of disc.**
Habitat: Among rock, sand and loose gravel, high intertidal zone to water 2,730' (828 m) deep.
Range: Alaska to southern California.
Notes: This small brittle star is very mobile and can often be found in tidepools. It broods its young. Its diet consists primarily of diatoms and detritus, and it can produce bioluminescence. Cells at the bases of the spines can be stimulated chemically in the laboratory setting to emit a glowing yellow-green luminescence. The significance of this is unknown at present.

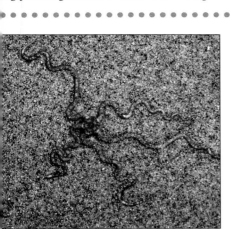

Long-armed Brittle Star

Amphiodia occidentalis

Other Names: Burrowing brittle star, long arm brittle star; formerly *Diamphiodia occidentalis*.
Description: Gray to tan with a dark spot on the disc at each arm. Blunt, flattened spines are present along the arms.
Size: Disc to $1/2"$ (1.3 cm) in diameter. **Length of arms 10 to 15 times diameter of disc** (to $6 5/8"/17$ cm).
Habitat: In sheltered sandy areas, mud flats and algal holdfasts from the low intertidal zone to water 1,210' (369 m) deep.
Range: Kodiak Island, Alaska, to San Diego, California.
Notes: The long-armed brittle star is a delicate and well-named creature. Its long arms are extremely fragile and easily fragmented. This species often burrows into sand or mud reaching shallow depths. With its tube feet moving up and down, it literally moves straight down, disappearing from view. The tips of the arms, however, are often not buried; they are used to feed at night.

SPINY-SKINNED ANIMALS

WESTERN SPINY BRITTLE STAR
Ophiothrix spiculata

OTHER NAMES: Spiny brittle star; sometimes misspelled *Ophiothryx spiculata*.
DESCRIPTION: Color ranges widely from brown or green to orange or yellow, with variable patterns. **Arms and disc look fuzzy** with numerous long, erect spines, each with many minute spinelets.
SIZE: Disc to 3/4" (1.9 cm) in diameter. Length of arms to 6" (15 cm) or 4 to 8 times diameter of disc.
HABITAT: Under rocks and among algal holdfasts from the mid-intertidal zone to water 6,794' (2,059 m) deep.
RANGE: Moss Beach, San Mateo County, California, to Bahía de Sechura, Peru.
NOTES: This elegant species lives in a truly remarkable range of depths. From where it clings, it extends a few arms to filter feed from the water, using a sticky mucus secretion. Larger food is also captured and transferred via the large podia (tube feet) to its jaws, in a co-ordinated fashion. At deep locations this star has been found to reach unbelievable populations, perhaps into the millions!

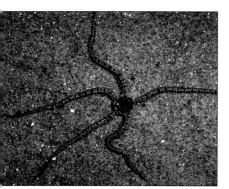

BANDED BRITTLE STAR
Ophionereis annulata

OTHER NAME: Ringed brittle star.
DESCRIPTION: Color varies from gray to light brown with **dark rings circling each arm**.
SIZE: Disc to 3/4" (1.9 cm) in diameter. Length of arms to 6 1/2" (16 cm), or 7 to 9 times diameter of disc.
HABITAT: In sand, under rocks and in tidepools from the mid-intertidal zone to 200' (60 m).
RANGE: San Pedro, Los Angeles County, California, to Esmeraldas, Ecuador, and the Galapagos Islands.
NOTES: The banded brittle star is an extremely fragile species that readily breaks off its arms if handled. It moves rapidly by way of the walking actions of its podia (tube feet) rather than the entire arms. Food consists of detritus and larger portions picked up by the tip of an arm; the arm is then coiled to bring the food to the mouth. Several banded brittle stars are often found together.

Flat-Spined Brittle Star

Ophiopteris papillosa

OTHER NAMES: Serpent star, blunt-spined brittle star, *Ophiopterus papillosa*, *Ophioptereis papillosa*.
DESCRIPTION: Arms are light brown, banded with dark brown. Disc is covered in flattened granules. **Spines are flattened and blunt.**
SIZE: Disc to about $1/2$" (1.2 cm) in diameter. Length of arms to approximately 2" (5 cm), or 3 to $4^{1/2}$ times diameter of disc.
HABITAT: Under rocks and in tidepools from the very low intertidal zone to water 460' (140 m) deep.
RANGE: Barkley Sound, BC, to Isla Cedros, Baja California, México.
NOTES: The flat-spined brittle star, like the western spiny brittle star (see p. 200), use mucus secretions to capture small food particles. Both species are also active carnivores, feeding on a variety of small creatures.

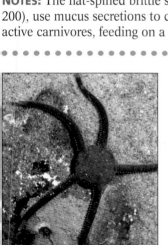

Smooth Brittle Star

Ophioplocus esmarki

OTHER NAMES: Esmark's serpent star, Esmark's brittle star.
DESCRIPTION: Color varies from red-brown to brown. Arms have several small plates and very short spines that can be folded flat against the arms. **Smooth surface** overall and a **large disc**.
SIZE: Disc to $1^{1/4}$" (3.1 cm) in diameter. **Length of arms** to $3^{3/4}$" (9.5 cm), **2 to 3 times diameter of disc**.
HABITAT: In sandy mud areas under rocks, in tidepools and among algal holdfasts, from the mid-intertidal zone to water 230' (70 m) deep.
RANGE: Tomales Bay, Marin County, California, to San Diego, California.
NOTES: The smooth brittle star is a locally common, rugged, slow-moving species that does not easily break its arms if handled. It fertilizes its eggs internally, then broods the eggs in pouches to develop directly into young in July. This distinctive brittle star scavenges on a variety of animal tissues.

SPINY-SKINNED ANIMALS

PANAMANIAN BRITTLE STAR
Ophioderma panamense

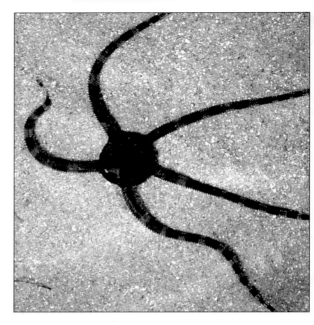

OTHER NAMES: Snakeskin brittle star, Panamanian serpent star, Panama brittle star, *O. panamensis*.
DESCRIPTION: Arms are olive to gray-brown with **light-colored bands in adults**, while juveniles are ivory-colored overall. **Disc is covered with tiny granules**. Short spines are present on the arms.
SIZE: Disc to 1 3/4" (4.5 cm) in diameter. **Arm length** to 10" (25 cm), or **3 to 6 times diameter of disc**.
HABITAT: Under rocks in tidepools and among algal holdfasts, from the midintertidal zone to water 130' (40 m) deep.
RANGE: Channel Islands, California, to Paita, Peru.
NOTES: The Panamanian brittle star is a large, fast-moving, warm-water species that is most active at night, scavenging along the seashore and feeding on a variety of small prey. This star has a total of 20 bursal or genital slits. Each slit opens into a pouch that plays a part in breathing, releasing gametes (sperm or eggs) and brooding its young.

URCHINS AND SAND DOLLARS
Class Echinoidea

There are more than 800 species of sea urchins and sand dollars worldwide. These animals feed with the help of a unique jaw-like apparatus known as Aristotle's lantern, an arrangement of parts, including teeth, with which the animal can eat tough seaweeds.

Sea urchins are covered with movable spines, which come in a wide range of colors and which can be blunt or sharp, long or short. Urchins make their way into our kitchens on occasion: their roe (see red sea urchin, p. 203) are eaten raw in sushi, and they are sautéed, added to omelettes or soups, and cooked in a variety of other ways.

Purple Sea Urchin
Strongylocentrotus purpuratus

Other Name: Purple spined sea urchin.
Description: Adults purple or purplish green, spherical and **covered with short, stout spines**. Length of spines is approximately a third the diameter of the round test (skeleton).
Size: Test to 4" (10 cm) wide, spines to 1" (2.5 cm) long.
Habitat: On exposed rocky shorelines, low intertidal zone to water 525' (160 m) deep.
Range: Cook Inlet, Alaska, to Baja California, México.
Notes: Large groups of this species often occur in the lower intertidal zone. Their chief food is seaweed. Some individuals have been found to live longer than 30 years. It is estimated that one female urchin can produce up to 20 million eggs in one year. Purple sea urchins can excavate holes in rock by using their sharp spines and teeth over time. While residing in their holes, they are protected from the pounding surf.

Red Sea Urchin
Strongylocentrotus franciscanus

Other Name: Giant red urchin.
Description: Red to purple test (skeleton) with **elongated spines**, similarly colored. Length of spines is normally half the diameter of the test.
Size: Test to $6^{1/2}$" (17 cm) wide, spines to 3" (7.6 cm) long.
Habitat: On exposed rocky shorelines, low intertidal zone to water 300' (91 m) deep.
Range: Alaska to Baja California, México.
Notes: This urchin is not as common intertidally as the purple sea urchin but can often be found in the same areas. The red urchin feeds on pieces of brown and red seaweed. Adults have been known to reach 20 years of age. The reproductive organs (roe) of the red sea urchin are harvested for shipment to Japan. It is a delicacy, eaten raw as a garnish for sushi and other dishes. The purple sea urchin (above) is a similar species, easily distinguished by its short spines.

SPINY-SKINNED ANIMALS

ECCENTRIC SAND DOLLAR
Dendraster excentricus

OTHER NAME: Pacific sand dollar.
DESCRIPTION: Covering of tiny spines and tube feet gives live individual a very dark, velvety, purple color; dead tests (shells) are off-white. 5-leafed pattern of tiny holes on dorsal side of test.
SIZE: To 3" (7.6 cm) in diameter.
HABITAT: On sandy beaches, low intertidal zone to water 50' (15 m) deep.
RANGE: Alaska to Baja California, México.
NOTES: In subtidal areas, density of sand dollars can be staggering—more than 523 per square yard (625 per m^2), a great deal more than are found intertidally. Sand dollars are unable to right themselves if overturned, and eventually die. Individuals living in intertidal areas are known to bury themselves, while those in quiet subtidal areas do not. In exposed subtidal areas, sand dollars lie flat, while in protected subtidal areas they rest at an angle to the waves. The sand dollar eats by moving diatoms and detritus to its central mouth with its tiny hair-like cilia. Individuals have been known to live 13 years.

A sand dollar test (shell).

SEA CUCUMBERS
Class Holothuroidea

Sea cucumbers are echinoderms (spiny-skinned animals) which are related to sea urchins. Most species have separate sexes. Reproduction occurs in two ways: some species produce many small eggs, which eventually develop into pelagic larvae; other species produce a few rather large yolk-filled eggs, which hatch directly into small cucumbers.

CALIFORNIA SEA CUCUMBER

Parastichopus californicus

OTHER NAMES: Giant sea cucumber, large red cucumber, common sea cucumber, California stichopus; formerly *Stichopus californicus*.
DESCRIPTION: Usually mottled reddish brown with soft projections along the entire body. Body is elongated and filled with water.
SIZE: To nearly 20" (50 cm) long, to 3" (7.6 cm) in diameter.
HABITAT: In exposed and sheltered areas, low intertidal zone to water 295' (90 m) deep.
RANGE: Gulf of Alaska to Baja California, México.
NOTES: The separate sexes of this sea cucumber mature at 4 years. The animal is harvested for the thin muscles running along the inside of the body wall. Its chief predator other than man is the sunflower star (see p. 197), to which it has been known to react in an interesting manner when in extreme danger. The internal organs are ejected from the anus, producing a sticky pile of viscera that could distract a predator. These organs are regenerated within 6–8 weeks.

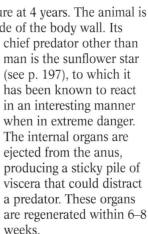

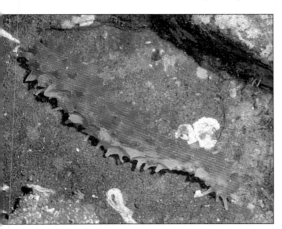

Young individual in a tidepool.

SPINY-SKINNED ANIMALS

WARTY SEA CUCUMBER
Parastichopus parvimensis

OTHER NAMES: Lesser California sea cucumber, southern California sea cucumber; formerly *Stichopus parvimensis*.
DESCRIPTION: Chestnut brown on the upper or dorsal side. Small papillae (**projections**) are **tipped with black**. Several larger "warty" projections may also be present on the dorsal surface. No tube feet are present on dorsal side.
SIZE: To 10" (25 cm) long.
HABITAT: In sandy or muddy areas from the low intertidal to 90' (27 m).
RANGE: Monterey Bay, California, to Punta San Bartolome, Baja California, México.
NOTES: The warty sea cucumber is a rapid mover in the world of sea cucumbers. It has been observed traveling 39" (1 m) in a mere 15 minutes. This species feeds on various organic materials from sediments. This cucumber is similar to the California sea cucumber (see p. 205), which grows to a larger size.

STIFF-FOOTED SEA CUCUMBER
Eupentacta quinquesemita

OTHER NAMES: White sea gherkin, white sea cucumber.
DESCRIPTION: **White** to cream-colored; shaped somewhat like a sausage with **rigid, spiny-looking tube feet**. 8 large branched tentacles and 2 small tentacles are located at one end of the body.

SIZE: To 3" (7.5 cm) long.
HABITAT: In crevices or between rocks, low intertidal to shallow subtidal zone.
RANGE: Sitka, Alaska, to Baja California, México.
NOTES: The enemies of this sea cucumber include the striped sun star (see p. 194) and sunflower star (p. 197). The six-rayed star (p. 195) feeds on juveniles.

SEA SQUIRTS
Phylum Urochordata

Lightbulb Tunicate
Clavelina huntsmani.

SEA SQUIRTS

There are three types of sea squirts (also called ascidians or tunicates): solitary, colonial and compound. All are covered with an exterior coating called a tunic, hence the name *tunicate*. Solitary species, often referred to as sea squirts, are oval, elongated or irregular in shape, and an individual is usually attached directly to the substrate by its side or a base. All sea squirts have 2 siphons, one to breathe and to obtain water for food, and the other to expel water and non-food particles. Colonial or social tunicates reproduce by budding or cloning to produce additional individuals from a single original member. Compound sea squirts are much different: many individuals (zooids) are packed together and form a fleshy common tunic. All individuals work together to ensure the survival of the compound organism.

It is truly remarkable that the simple-looking tunicates are one of the most advanced group of organisms found in the intertidal zone. Animals in the phylum Urochordata have a primitive nerve cord (notochord), and they are distantly related to fish, whales and man.

A unique feature of tunicates is the heart, which reverses its beating every few minutes to change the direction in the flow of blood. Any advantage this system provides is currently unknown.

SPINY-HEADED SEA SQUIRT
Boltenia villosa

OTHER NAMES: Spiny-headed tunicate, hairy sea squirt, stalked hairy sea squirt, bristly tunicate.
DESCRIPTION: Red, orange or tan covering with orange to deep red siphons. Spines cover the body, which may or may not be on an elongated stalk.
SIZE: To 4" (10 cm) high.
HABITAT: On hard substances, low intertidal zone to water 330' (100 m) deep.
RANGE: Prince Rupert, BC, to San Diego, California.
NOTES: This sea squirt and the long-stalked sea squirt (see p. 209) have been found to separate vanadium (a metal element used in manufacture of various alloys) from their environment and to concentrate it in their bodies. Several sea stars, including the leather star (see p. 194), have been known to prey on the spiny-headed sea squirt.

LONG-STALKED SEA SQUIRT

Styela montereyensis

OTHER NAMES: Stalked tunicate, Monterey stalked tunicate.
DESCRIPTION: Thick, leathery reddish orange covering, with several distinctive elongated wrinkles.
SIZE: To 10" (25 cm) high.
HABITAT: On protected and exposed rocky shores, low intertidal zone to water 100' (30 m) deep.
RANGE: Vancouver Island, BC, to Baja California, México.
NOTES: Long-stalked sea squirts inhabiting quiet bays grow to a larger size than those in exposed locations. One individual has been known to reach 9" (23 cm) in height in 3 years. The top of this tunicate may be pointed straight up or its weight may pull it downward at right angles to the stalk. This species can absorb and store vanadium from its environment (see spiny-headed sea squirt, p. 208).

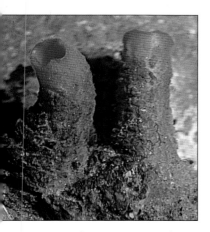

WARTY SEA SQUIRT

Pyura haustor

OTHER NAMES: Warty tunicate, solitary sea squirt, wrinkled seapump.
DESCRIPTION: The 2 exposed red siphons are often the only recognizable part of this species. Globular body is normally covered with debris and encrusting organisms.
SIZE: To 3" (7.6 cm) wide.
HABITAT: Attached to hard substrates, low intertidal zone to depths of 660' (200 m).
RANGE: Aleutian Islands, Alaska, to San Diego, California.
NOTES: The warty sea squirt reproduces at all times of the year, when both eggs and sperm are released into the ocean. Like all tunicates, it is a filter feeder. The striped sun star (see p. 194) is known to prey on this tunicate.

SEA SQUIRTS

LIGHTBULB TUNICATE
Clavelina huntsmani

OTHER NAME: Lightbulb ascidian.
DESCRIPTION: The elongated **tunic is transparent with pink to yellow lines** visible inside.
SIZE: Individuals to 2" (5.1 cm) long, 3/8" (1 cm) wide. Colonies can reach 8" (20 cm) across.

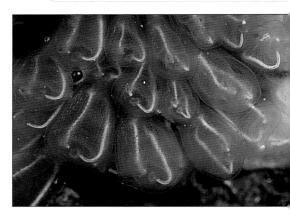

HABITAT: On rocks and ledges from the low intertidal zone to water 100' (30 m) deep.
RANGE: Vancouver Island, BC, to San Diego, California.
NOTES: The lightbulb tunicate is an aptly named colonial ascidian with a clear outer tunic and colored lines resembling the filaments of a light bulb. Most commonly found in spring and summer, this sea squirt buds asexually to produce additional individuals to make up a colony. Bright orange embryos are also produced and brooded in a cavity within the tunic. The outer clear tunic appears very fragile but a gentle touch will reveal how tough it is.

CALIFORNIA SEA PORK
Aplidium californicum

OTHER NAMES: Sea pork; formerly *Amaroucium californicum*.
DESCRIPTION: Color of the **colony** varies from **yellowish** to gray or brown. The **surface of this encrusting species is smooth, slick** and not normally covered with sand.
SIZE: Colony to 8" (20 cm) across and 1 1/4" (3 cm) thick.
HABITAT: On rock from the mid-intertidal zone to water 276' (84 m) deep.
RANGE: Alaska to La Paz, Baja California Sur, México.
NOTES: Sea pork, a very common colonial tunicate, is found in the shade of rocks in areas protected from the surf. Its name comes from its color and slab-like shape, which cause it to resemble pork fat. Enemies of this species include the leather star (see p. 194) and bat star (p. 193).

Red Sea Pork
Aplidium solidum

OTHER NAME: Red ascidian; formerly *Amaroucium solidum*.
DESCRIPTION: Color varies from bright red to pink or orange-brown; species gathers in massive, **irregular, fleshy forms**. The outer layer is tougher than the interior.
SIZE: Colony to 8" (20 cm) across, 2" (5 cm) thick, but smaller colonies are more common.
HABITAT: On rocks and similar objects, low intertidal zone to water 50' (15 m) and occasionally to 132' (40 m) deep.
RANGE: Vancouver Island, BC, to San Diego, California.
NOTES: Red sea pork is found in sheltered areas where a current is present. It is a common ascidian, most abundant in spring and summer but present year-round. The opalescent nudibranch (see p. 124) has been observed feeding on this species.

Yellow Lobed Tunicate
Eudistoma ritteri

OTHER NAMES: Clubbed sea squirt; formerly *Archidistoma ritteri*.
DESCRIPTION: Colony **yellowish** to whitish and shaped in irregular, **gelatinous lobes**. These lobes are smooth and firm in nature. Colonies also form on rocks, as sheets with mounds or finger-like lobes.
SIZE: Colony to 1 1/2" (4 cm) high.
HABITAT: On rock from the low intertidal zone to water 66' (20 m) deep.
RANGE: Vancouver Island, BC, to San Diego, California.
NOTES: This sea squirt is found in areas with strong water currents but not in high-surf areas. The tunic contains strong sulfuric acid. Reproduction alternates between the sexual stage, when tadpole larvae are released during summer, and asexual (budding) stage, which takes place over the winter months.

SEA SQUIRTS

MUSHROOM TUNICATE
Distaplia occidentalis

OTHER NAMES: Club tunicate, western distaplia.
DESCRIPTION: Extremely variable in color, ranging from white to orange or purple.
SIZE: Colonies to 5" (12 cm) in diameter. Several colonies may grow together in the same area.
HABITAT: Low intertidal zone to subtidal waters 50' (15 m) deep. Often found on floats.
RANGE: Alaska to San Diego, California.
NOTES: The mushroom tunicate is a colonial or social tunicate made up of several individuals. It gets its name from the shape of small colonies when under water. Each member of the colony is called a zooid, and together these zooids produce an oral siphon to admit water into that colony. Smaller, separate siphons pass the water out of the body, once the food has been filtered out. It is believed that this tunicate's high acid content may help protect it from its enemies.

Purple form with closed siphons.

Orange form with open siphons.

STALKED COMPOUND TUNICATE
Distaplia smithi

OTHER NAMES: Paddle ascidian, Smith's distaplia, club tunicate.
DESCRIPTION: Color varies from cream to orange-brown. Colonies consist of clusters of paddle-shaped lobes, each attached with a stalk but only visible when viewed underwater. Out of water, the colony looks like tapioca pudding.
SIZE: Stalks to 2" (5 cm) long and colony normally to 2" (5 cm) across.
HABITAT: Low intertidal zone to water 50' (15 m) deep.
RANGE: Prince William Sound, Alaska, to Monterey County, California.
NOTES: When this tunicate is out of water, several of its characteristics cannot be observed. This species is never found encrusted with sand, and it favors areas with a strong current. To reproduce, zooids (individuals) release sperm prior to the production of eggs—a type of reproduction (called "protrandrous") that is uncommon in tunicates.

Multi-Lobed Tunicate
Ritterella aequalisiphonis

OTHER NAMES: Formerly *Amaroucium aequalisiphonis, Sigillinaria aequali-siphonis*.
DESCRIPTION: Orange-brown in color. Each colony appears as a compact mound with a hemispherical shape. This is a **sand-encrusted colonial species**.
SIZE: Individual lobes to 1/4" (7 mm) wide and cluster to 3" (8 cm) across.
HABITAT: Sheltered sites with moderate surf, from the low intertidal zone to water 50' (15 m) deep.
RANGE: Puget Sound, Washington, to San Diego, California.
NOTES: The multi-lobed tunicate forms several hemispherical shapes together that make up a larger colony. Several species of crustaceans are often found in the sheltered areas between the mounds.

Orange Chain Tunicate
Botrylloides diegensis

DESCRIPTION: Color of test (shell) is orange to pinkish yellow, with a purple or brown ground color. Flat, irregular sheets are formed where the zooids (individuals) are arranged in distinctive, elongated groups, giving a chain-like appearance to the colony.
SIZE: Colony to 6" (30 cm) across and 3/16" (5 mm) thick.
HABITAT: Low intertidal zone to shallow subtidal waters.
RANGE: Throughout California.
NOTES: The orange chain tunicate is a compound species that occurs in a distinctive meandering, chain-like pattern of zooids that form an irregular encrusting sheet. The tests are often bright orange, suppressing the ground color so that the colony appears to be orange overall. Tiny larvae are brooded in special incubating pouches on the sides of the zooids.

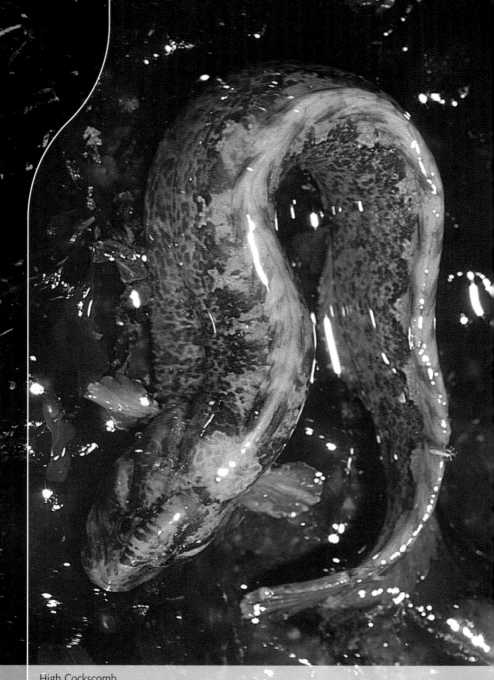

High Cockscomb
Anoplarchus purpurescens.

FISHES
Phylum Chordata

Fish can be found in all marine habitats from tidepools to deep subtidal waters. It is one of our more important foods, and various species have been harvested by man for centuries; but although a large number of species occur in the intertidal habitat—most of them small and hardy individuals—few intertidal species have been harvested.

SWELL SHARK (EGG CASE)

Cephaloscyllium ventriosum

OTHER NAMES: Swellshark; formerly *Cephaloscyllium uter*.

DESCRIPTION: The **egg case** is a smooth purse-shaped capsule that holds a single egg. When fresh, the egg case is greenish amber in color, but eventually it dries to black.

SIZE: Egg case 4x1 1/4" (10x3.1 cm).

HABITAT: Often found washed up on exposed sandy beaches.

RANGE: Monterey Bay, California, to southern México and central Chile.

NOTES: The swell shark gets its name from its habit of swallowing water or air to swell up when distressed. The eggs are laid in seaweed, where curly tendrils attach to kelp in order to help the eggs remain stable for 7–9 months before hatching. Empty egg cases are often found washed up on the beach. This common shark is a member of the cat-shark family with somewhat cat-like features, especially the eyes. The adult body is spotted with saddle-like patterns over the back.

FISHES

NORTHERN CLINGFISH
Gobiesox maeandricus

OTHER NAME: Flathead clingfish.
DESCRIPTION: Color varies from light olive to red mottled with dark brown. Distinctive flat head tapers quickly to tail. **Belly has large circular adhesive disc.**
SIZE: To 6 1/2" (16 cm) long.

HABITAT: Under rocks, often where tidal currents are strong, intertidal to subtidal zones.
RANGE: Southeastern Alaska to México.
NOTES: This fish usually uses its adhesive disc to cling to the underside of a rock, where it waits for food to arrive. Its diet includes the white-lined chiton, lined chiton, shield limpet, plate limpet, ribbed limpet, mask limpet and red rock crab. This fish is also cannibalistic, feeding on sculpins and smaller northern clingfish. If you happen to find this species hiding under a rock, be sure to return the rock to its original position. Many types of invertebrates also make this habitat their home. Another species, the California clingfish *Gobiesox rhessodon*, is also found from Pismo Beach to central Baja California, México. It is a smaller fish with a large fleshy pad in front of each pectoral fin and several dark markings radiating from its eyes.

TIDEPOOL SCULPIN
Oligocottus maculosus

DESCRIPTION: Color varies from brown to green or reddish on dorsal side, fading to lighter hues on underside.
SIZE: To 3 1/2" (8.8 cm) long.
HABITAT: In tidepools, in nearly any intertidal zone.
RANGE: Bering Sea to southern California.
NOTES: The tidepool sculpin has been well studied by biologists. They have found this species very tolerant of extreme changes in temperature, from the heat of tiny tidepools in direct sun to the cool waters of high tide. This sculpin is able to use its sense of smell to "home" back to its original tidepool if displaced. It is also capable of changing its color to match its environment.

CALIFORNIA GRUNION
Leuresthes tenuis

OTHER NAME: Grunion.
DESCRIPTION: Color is greenish above, silvery below with a blue to violet midline. Body is elongated, with a spiny first dorsal fin located above the anal fin.
SIZE: To 7$^{1/2}$" (19 cm) long.
HABITAT: On sandy beaches to 60' (18 m).
RANGE: San Francisco, California, to Bahía de San Juanico, Baja California, México.
NOTES: The California grunion is famous for spawning in the sandy beaches of southern California. At night, from March through September, females burrow backwards into the sand to lay their eggs just after the highest tides. These runs are predictable, occurring on the second through fourth nights after both the full and new moons. Each female is accompanied by one or more males, which also come ashore to fertilize the eggs. The eggs remain in the warm, moist sand to incubate and hatch with the next series of spring tides. Limited harvesting has been allowed under strict regulations. Contact the California Department of Fish and Game for current regulations.

TIDEPOOL SNAILFISH
Liparis florae

OTHER NAME: Shore liparid.
DESCRIPTION: Color is uniform but varies considerably from brown to green, purple or yellow. This fish is noted for its loose skin. **Small adhesive disc on belly** acts as a suction disc to aid the fish in attaching to the underside of rocks.
SIZE: To 5" (12.7 cm) long.
HABITAT: In rocky areas, often among algae or surf-grasses, or attached to the underside of large rocks; low intertidal zone to shallow rocky depths.
RANGE: Bering Sea, Alaska, to Point Conception, Santa Barbara County, California.
NOTES: The tidepool snailfish can change its color somewhat, turning lighter or darker to match its environment. Its eyes are noticeably small for its size. It feeds primarily on shrimp and similar animals living in its rocky habitat.

HIGH COCKSCOMB
Anoplarchus purpurescens
OTHER NAMES: Cockscomb prickleback, crested blenny.
DESCRIPTION: Color varies from brown to olive, purple, orange or black. **Fleshy crest on top of head**. Light bar normally present in front of tail fin; no pelvic fins.
SIZE: To 7 3/4" (20 cm) long.
HABITAT: In tidepools and under rocks, intertidal zone.
RANGE: Aleutian Islands, Alaska, to Santa Rosa Island, California.
NOTES: The beachcomber is most likely to find this fish under an intertidal rock, where several individuals are often grouped together. In the darkness of night, the high cockscomb comes out from hiding to feed. Green algae, worms, mollusks and crustaceans are important foods. Females grow to be larger than males. They are known to release 2,700 eggs, which are then guarded and fanned by the female as she wraps her body around them. The common garter snake is a predator of this species.

MOSSHEAD WARBONNET
Chirolophis nugator
OTHER NAMES: Mosshead prickleback, ornamental blenny.
DESCRIPTION: Overall color varies from reddish brown to orange-brown. The top of its head is graced with many cirri (thread-like projections), which form a flattened crest-like covering. A total of 13 eye-spots are also found along the dorsal fin of the male, while bars are present in females.
SIZE: To 6" (15 cm) long.
HABITAT: Under and between rocks from the very low intertidal to 265' (80 m).
RANGE: Kodiak Island, Alaska, to San Miguel Island, California.
NOTES: The mosshead warbonnet is a striking fish to discover at the lowest of tides, and its distinctive headdress makes it easy to identify. It is sometimes found hiding in shells or rock crevices with only its head showing. Spawning occurs from late winter to early spring. It is a known predator of nudibranchs.

Black Prickleback

Xiphister atropurpureus

OTHER NAME: *Epigeichthys atropurpureus*.
DESCRIPTION: Color varies from **reddish brown to black**. **2 dark bands with light edges radiate backward from the eyes**. The dorsal fin joins with the tail and anal fins. A conspicuous **light band is present at base of tail fin**.
SIZE: To 12" (30 cm) long.
HABITAT: Under rocks on exposed shorelines, low intertidal zone.
RANGE: Kodiak Island, Alaska, to northern Baja California, México.
NOTES: The black prickleback feeds on an assortment of algae in addition to a few species of small invertebrates. It breeds in late winter and spring. The egg mass is typically laid under a rock or similar object, then guarded by the male for about 3 weeks before the eggs hatch. This fish is most often found beneath a rock in little or no water. These damp conditions help keep the fish and other organisms from drying out, so if you turn over any rocks, please replace them carefully.

Crescent Gunnel

Pholis laeta

OTHER NAME: Bracketed blenny.
DESCRIPTION: Lime-colored on top, lighter below. **Series of crescent-shaped markings along the back**, just below long dorsal fin.
SIZE: To 10" (25 cm) long.
HABITAT: Under seaweed or rocks and in tidepools, mid-intertidal zone to water 240' (73 m) deep.
RANGE: Bering Sea to northern California.
NOTES: This wriggling fish feeds on a variety of small bivalves and nips off the cirri (feeding appendages) of barnacles. It wraps its body around its eggs to protect them from predators until they hatch. The crescent gunnel is a common species that has been known to live as long as 6 years. Mergansers are likely important predators.

FISHES

SADDLEBACK GUNNEL
Pholis ornata

OTHER NAME: Saddled blenny.
DESCRIPTION: Green to brown with elongated body and elongated dorsal and anal fins. **Repeated U-shaped arch pattern along upper body.**
SIZE: To 12" (30 cm) long.
HABITAT: Under rocks or with eel-grass, intertidal zone to subtidal waters 120' (37 m) deep.
RANGE: Bering Sea to central California.
NOTES: The gunnels are a group of fish also frequently referred to as blennies. "Blenny" is a European term for several families of fish that include gunnels, pricklebacks and others. The eggs of the saddleback gunnel are laid in late winter to the first hint of spring and guarded by both parents. The crescent gunnel (p. 219) is similar-looking, with crescent-shaped marks along its upper body.

BLIND GOBY
Typhlogobius californiensis

DESCRIPTION: Flesh-colored body with darker cheeks. No eyes are present on the adults. The pelvic fins are fused into one.
SIZE: To 2 1/2" (6.3 cm) long.
HABITAT: Intertidal zone to water 25' (8 m) deep.
RANGE: San Simeon, California, to Bahía Magdalena, Baja California Sur, México.
NOTES: The blind goby lives in the burrows of the bay ghost shrimp (see p. 171). Its pink color is a direct result of living in an oxygen-deficient environment: the blood vessel network, just below the skin's surface, is greatly increased to help the fish obtain enough oxygen. One pair of blind gobies live in a burrow. There the female lays up to 15,000 eggs, which eventually hatch into young that have functional eyes. But the eyes are covered over within about 6 months.

Sea Sacs *Halosaccion glandiforme*.

SEAWEEDS
Phyla Chlorophyta, Phaeophyta and Rhodophyta

SEAWEEDS

Seaweeds or marine algae use pigments such as chlorophyll to trap energy from the sun and store it as chemical energy in the bonds of a simple sugar (glucose). These plants are an important part of both the intertidal and subtidal ecosystems. Their presence provides food and shelter, as well as oxygen, to the wide variety of invertebrates and vertebrates found here. They are the base of many food webs to which all species are linked. Seaweeds are classified into 3 separate groups (phyla) according to the types of pigments present: green algae (phylum Chlorophyta), brown algae (phylum Phaeophyta) and red algae (phylum Rhodophyta).

People all over the world have gathered and eaten seaweeds, probably for hundreds of years. At least 70 species have been harvested from the Pacific Ocean.

GREEN ALGAE
Phylum Chlorophyta

Green algae contain chlorophylls a (a primary photosynthetic pigment in plants that release oxygen) and b (an accessory pigment also found in vascular plants). In this way they are very similar to green land plants. The green algae are found in shallow waters where chlorophyll is most efficient. Green algae store their sugars as starch.

TUBEWEED
Enteromorpha sp.

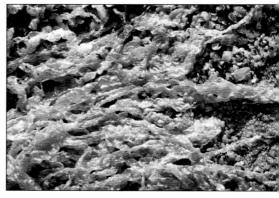

OTHER NAMES: Sea hair, link confetti; formerly *Ulva* sp.
DESCRIPTION: Bright yellowish green with elongated, unbranched **hollow** blades.
SIZE: Normally to 8" (20 cm) long, occasionally to 3.3' (1 m).
HABITAT: In rocky areas, upper to mid-intertidal zone and at freshwater seeps. Also an epiphyte on other species of algae (attached to the organism with no harm to it), often in tidepools.
RANGE: Aleutian Islands, Alaska, to México.
NOTES: Tubeweed is often bleached white by the sun. Upon close inspection, its fine tubes resemble hairs. This seaweed has been used in stews: a small amount flavours a large volume of stew. This species is similar to some sea lettuces (see opposite).

Sea Lettuce
Ulva fenestrata

OTHER NAME: Window seaweed.
DESCRIPTION: Light green or darker oblong blade. Smooth surface, often split into broad lobes, with ruffled edge. Several small holes may be present on blades.
SIZE: To 7" (18 cm) long.
HABITAT: Attached to rocks or other algae species, all intertidal zones. Also found floating on mud flats, lower intertidal.
RANGE: Bering Sea to Chile.
NOTES: This alga is very tolerant of a great range of temperatures. It is often seen out of water, somewhat dried out, when the tide is low. Under fertile conditions, it will cover large areas. Similar species of sea lettuce have traditionally been eaten by Hawaiians, mixed with other seaweeds and served with sushi, made into a light soup and made into stews, and used in a variety of other ways.

Corkscrew Sea Lettuce
Ulva taeniata

OTHER NAME: Formerly *Ulva fasciata* var. *taeniata*.
DESCRIPTION: Bright green **blades elongated** and ruffled **with a spiral twist**. Blades can be either unbranched or branched.
SIZE: To 20" (50 cm) tall and $1^{1/2}$" (4 cm) broad.
HABITAT: On rocks from mid- to low intertidal zone.
RANGE: Southern BC to Baja California, México.
NOTES: This common green alga is thin: only 2 cells thick. When viewed underwater, its corkscrew pattern is especially visible. It is a widespread species, present in Peru, New Zealand and Australia as well as the west coast of North America.

SEAWEEDS

TANGLE WEED
Acrosiphonia coalita

OTHER NAMES: **Gametophyte stage:** green rope; formerly *Spongomorpha coalita*; **Sporophyte stage:** formerly *Chlorochytrium inclusum*.
DESCRIPTION: Occurs in 2 separate stages. **Gametophyte stage:** bright green with many filaments, which become entangled to resemble rope. **Sporophyte stage:** tiny greenish spots on red algae.
SIZE: To 12" (30 cm) tall.
HABITAT: On exposed rocky areas, mid- to low intertidal zone.
RANGE: Alaska to San Luis Obispo County, California.
NOTES: The rope-like strands of tangle weed occur only in the gametophyte (sexual) stage of its life cycle. A spore (saprophyte) produced at this stage lives in the crustose stage of a few species of red algae, including papillate seaweed (see p. 246). Tangle weed is sometimes found as an epiphyte, attached to another species of algae without harming it. The sporophyte (asexual) stage is very difficult to find: it is unicellular and present only in the tissue of various red algae species.

SEA MOSS
Cladophora sp.

DESCRIPTION: Bright green with a filamentous shape, growing in low mats.
SIZE: To 2" (5 cm) high.
HABITAT: In rocky areas, mid- to low intertidal zone.
RANGE: Alaska to Baja California, México.
NOTES: The structure of sea moss enables it to hold large volumes of water, preventing it from drying out when the tide recedes. There are a few different species of sea moss, all growing low to the ground and forming moss-like mats or tufts.

Sea Pearls
Derbesia marina

OTHER NAMES: Green sea grape; **Globose stage:** Formerly *Halicystis ovalis*.
DESCRIPTION: Light to dark green. Occurs in 2 separate stages. **Globose stage:** greenish to black spherical globules, especially on encrusting coralline algae (see p. 240). **Filamentous stage:** irregularly branched elongated filaments.
SIZE: Globose stage: to $3/8$" (1 cm) in diameter. **Filamentous stage:** to $1 1/2$" (4 cm) long.
HABITAT: On encrusting coralline algae, sponges and rocks, in exposed and sheltered situations, low intertidal zone to water 33' (10 m) deep.
RANGE: Aleutian Islands, Alaska, to La Jolla, San Diego County, California.
NOTES: Only after careful laboratory study were the 2 distinct stages of this species determined to belong the same marine alga. The globose stage, often referred to as the "Halicystis" stage, is the sexual stage that produces gametes, while the filamentous stage reproduces by cell division.

Sea Staghorn
Codium fragile

OTHER NAMES: Dead man's fingers, sponge seaweed, felty fingers, green sea velvet.
DESCRIPTION: Dark green to blackish green with many thin, spongy cylindrical branches rising from a basal disc.
SIZE: To 16" (40 cm) high.
HABITAT: Attached to rocks, mid-intertidal to upper subtidal zone. Also often found in large tidepools.
RANGE: Alaska to Baja California, México.
NOTES: This seaweed resembles a sponge, but there are no green sponges found on the Pacific coast. The sea staghorn is very rich in vitamins and minerals, so it is often used in soups although it is hard to clean. It has invaded the New England coast, where it has caused some problems with the shellfish industry: the young plants often start growing on oysters, mussels and scallops, increasing the drag and causing the mollusks to be taken out to sea during storms. A small red alga, staghorn fringe *Ceramium codicola*, lives on sea staghorn and is not found anywhere else.

SEAWEEDS

SPONGY CUSHION
Codium setchellii

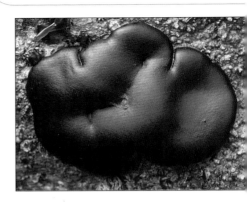

DESCRIPTION: Dark green to black, forming smooth, flat, irregular cushions. Texture changes from firm to spongy as the individual ages.
SIZE: To 5/8" (1.5 cm) thick, 10" (25 cm) across; smaller specimens are more common.
HABITAT: In very exposed rocky areas, often attached to rocks in sandy areas, low intertidal zone.
RANGE: Sitka, Alaska, to Baja California, México.
NOTES: This common, distinctive seaweed is sometimes covered by sand for extended periods.

BROWN ALGAE
Phylum Phaeophyta

Brown algae contain chlorophyll, but the green color is hidden by gold and brown pigments, which look light green to dark black because of the proportions of pigments. Several large brown algae are often referred to as kelp.

SEA CAULIFLOWER
Leathesia difformis

OTHER NAMES: Sea potato, *Tremulla difformis*.
DESCRIPTION: Yellow, roughly spherical, convoluted. Hollow and spongy.
SIZE: To 5" (12 cm) in diameter.
HABITAT: In exposed and protected rocky areas, all intertidal zones. Also epiphytic: growing on other species of algae with no harm to the other algae.
RANGE: Bering Sea to Baja California, México.
NOTES: When this species was first described by the famous botanist Carolus Linnaeus, it was thought to be a jelly fungus because of its unusual shape. The sea cauliflower is solid when young but becomes hollow as it matures. It is found in many parts of the world including Europe, Chile and Sweden.

Convoluted Sea Fungus

Petrospongium rugosum

OTHER NAME: Formerly *Cylindrocarpus rugosus*.
DESCRIPTION: Color varies from chestnut brown to dark brown. Entire plant is gelatinous and roughly **circular** with irregular **convoluted ridges** on the upper surface.
SIZE: To 3" (8 cm) in diameter.
HABITAT: On rocks in the high intertidal zone.
RANGE: Vancouver Island, BC, to Baja California, México.
NOTES: This distinctive alga somewhat resembles a hollow fungus. Patches of it can be found where conditions are favorable, often in late autumn. The species also occurs in Japan.

Banded Tidepool Fan

Zonaria farlowii

DESCRIPTION: Color varies from dark green to olive-brown. Normally in dense bush-like clusters. **The end of the thallus (blade) is fan-shaped** with multiple splitting. Concentric lines or bands radiate from the holdfast area.
SIZE: To 6" (15 cm) tall.
HABITAT: On sand-covered rocks in the mid-intertidal zone to water 65' (20 m) deep.
RANGE: Santa Barbara, California, to Isla Magdalena, Baja California, México.
NOTES: The banded tidepool fan is a member of the order Dictyotales, a group of brown algae most of which are found in tropical waters. The reproductive gametes are produced on separate plants. This common seaweed is very tolerant of being covered by sand in the tidepools of southern California.

SEAWEEDS

FLAT ACID KELP
Desmarestia ligulata

OTHER NAMES: Acid seaweed, wide desmarestia, flattened acid kelp; formerly *Fucus ligatulus*.
DESCRIPTION: Yellowish brown to dark brown with numerous branches covered in fine hairs. Flat blades are attached to a central stipe (stalk).
SIZE: Normally to 31" (80 cm) long, but has been known to reach 26' (8 m).
HABITAT: In rocky areas, low intertidal to upper subtidal zone.
RANGE: Alaska to South America.
NOTES: This species secretes sulfuric acid, which can damage its own tissue as well as other seaweeds. This causes dramatic color changes. Flat acid kelp often becomes established in areas where the giant perennial kelp (see p. 234) is hard hit by winter storms.

SPLIT KELP
Laminaria setchellii

DESCRIPTION: Rich brown to black with prominent stipe (stalk). Blades are split deeply, almost to the ends.
SIZE: Blades to 32" (80 cm) long, 10" (25 cm) wide.
HABITAT: In exposed rocky areas, low intertidal to upper subtidal zone.
RANGE: Vancouver Island, BC, to Ventura County, California.
NOTES: The stipe is stiff and stands erect in exposed, surf-swept stretches of the coast. Often this species is found in high enough concentrations to be called an "underwater forest." It is held in place with a very compact holdfast.

Oar Weed

Laminaria sinclairii

OTHER NAMES: Dense clumped laminarian; formerly *Lessonia sinclairii*.
DESCRIPTION: This kelp is a rich dark brown color with smooth, elongated blades. **Blades and stipe are slender and very slimy or glue-like.** Several stipes originate from the distinctive **creeping rhizome** (root-like stem).
SIZE: To 39" (1 m) long; 1 1/4" (3 cm) wide.
HABITAT: On rocks in the low intertidal zone; often in **sandy** areas.
RANGE: Hope Island, BC, to Santa Barbara County, California.
NOTES: Dense clumps of oar weed are commonly found attached to rocks at sandy sites. *Laminaria* species were once harvested for their iodine content. Since then, commercial harvesting has shifted to the production of algin (see giant perennial kelp, p. 234). Today these algae are being farmed as food.

Sea Cabbage

Hedophyllum sessile

OTHER NAME: Formerly *Laminaria sessilis*.
DESCRIPTION: Brown and leather-like with wrinkled blades attached to rock with a sturdy holdfast. Blades become smooth and deeply split in surf-swept areas.
SIZE: Normally to 20" (50 cm) long; occasionally to 5' (1.5 m) long, 32" (80 cm) wide.
HABITAT: On rock, mid-intertidal to upper subtidal.
RANGE: Aleutian Islands, Alaska, to Monterey County, California.
NOTES: This variable species grows completely different in different habitats. Rough waters give it a smooth form, while corrugated blades are found in quiet bays and similar habitats. The black Katy chiton (see p. 70) feeds on the sea cabbage. Sea urchins do not feed on it, so it can become very abundant in areas where urchins are present.

Smooth form.

Wrinkled form.

SEAWEEDS

SEERSUCKER
Costaria costata

OTHER NAMES: Five-ribbed kelp, ribbed kelp.
DESCRIPTION: Single brown to chocolate-brown blade with 5–7 prominent parallel ribs attached to large stipe (stalk).
SIZE: Blade 20–100" (50–250 cm) long, 4–12" (10–30 cm) wide.
HABITAT: In rocky areas and occasionally on wood, low intertidal to upper subtidal zone.
RANGE: Unalaska Island, Alaska, to San Pedro, California.
NOTES: Seersucker is found in sheltered locations and also on the open coast, where it grows somewhat narrower. Tattered and sometimes discolored blades are commonly seen in early summer. Sea urchins feed readily on this kelp.

OLD GROWTH KELP
Pterygophora californica

DESCRIPTION: Dark brown with long, woody stipe (stalk). A series of elongated blades originate near the tip.
SIZE: Stipe to $6^{1/2}$' (2 m) long, blades to 3' (90 cm) long.
HABITAT: In rocky areas, low intertidal zone to water 30' (9 m) deep.
RANGE: Cook Inlet, Alaska, to Baja California, México.
NOTES: Old growth kelp has been known to live to 20 years, with new blades growing each year. Its age can be calculated by counting the growth rings of its woody stipe.

Winged Kelp
Alaria marginata

Other Names: Alaria, ribbon kelp.
Description: Olive brown to rich brown with prominent linear blade on stipe (stalk). Smaller, wing-like blades (sporophylls) are attached to the base of the midrib or stipe, forming a cluster. These produce spores.
Size: Blade 8–10' (2.5–3 m) long, 6–8" (15–20 cm) wide.
Habitat: In rocky areas, often in areas exposed to high surf; low intertidal to upper subtidal zone.
Range: Kodiak Island, Alaska, to Point Conception, Santa Barbara County, California.
Notes: Winged kelp is dried and used in cooking soups and stews, as a substitute for kombu (the Oriental species *Laminaria japonica*). It is also deep fried and eaten like potato chips, and the midrib can be eaten fresh in salads, or added to spaghetti sauce. This is truly a versatile seaweed!

Feather Boa Kelp
Egregia menziesii

Other Names: Venus's girdle; formerly *Egregia laevigata, Egregia planifolia*.
Description: Color varies from olive green to brown. Irregular branches along entire stipe (stalk), and spherical floats often located sparsely along entire branch. Plant is held securely to rock by a holdfast.
Size: Branches to 33' (10 m) long, 14" (35 cm) wide.
Habitat: In rocky exposed areas, low intertidal to upper subtidal zones.
Range: Queen Charlotte Islands, BC, to Punta Eugenio, Baja California Sur, México.
Notes: This species has been used for many years as a fertilizer by coastal farmers. The very specialized seaweed limpet (see p. 80) can sometimes be found feeding on the blades of this seaweed, if you look closely.

SEAWEEDS

SEA PALM
Postelsia palmaeformis

DESCRIPTION: Olive brown, resembles a miniature palm tree. **Main stipe (stalk) round** and supports as many as 100 flattened, deeply grooved blades at the tip.
SIZE: To 24" (60 cm) high, blades to 9" (24 cm) long.
HABITAT: In areas exposed to heavy surf, mid- to lower intertidal zone.
RANGE: Hope Island, BC, to San Luis Obispo County, California.
NOTES: This very robust species takes a daily pounding of heavy surf, bouncing back like an elastic band with every wave. The sea palm is an annual, producing spores that germinate close to the parent plant. The shield limpet (see p. 75) can often be found living on the stipe.

DOUBLE POMPOM KELP
Eisenia arborea

OTHER NAMES: Southern sea palm, forked kelp.
DESCRIPTION: Color varies from dark brown to almost black. The stout, woody **stipe (stalk) is flattened, erect and forked** at its upper end. Blades are toothed and corrugated on the surface, and they grow from the forked ends of the flat stipe.
SIZE: Normally to 2' (60 cm) tall intertidally and 6½' (2 m) subtidally.
HABITAT: On exposed solid rock from the low intertidal zone to 108' (33 m).
RANGE: Queen Charlotte Islands, BC, to Isla Magdalena, Baja California, México; Japan.
NOTES: Although the range of the southern sea palm extends to BC, this alga is common only in southern California. It is a tough species that looks somewhat like a palm tree, hence its name. This kelp has been aged at more than 75 years by counting the growth rings of its woody central stipe.

Netted Blade
Dictyoneurum californicum

Description: Color varies from light to dark yellowish brown. Two flattened, equal stipes (stalks) grow from a shoe-like holdfast. **Blades are tapered on both ends with reticulate (net-like) veins** and no midrib.
Size: Normally to 6 1/2' (2 m) and occasionally to 13' (4 m) tall.
Habitat: From the low intertidal zone to water 33' (10 m) deep.
Range: Vancouver Island, BC, to Point Conception and the Channel Islands, California.
Notes: A lover of surf areas, netted blade prefers exposed coastal areas. Clumps of 25 to 100 blades are often found grouped together. Starting in mid-summer, darker reproductive structures grow between net-like veins. The base of this alga often harbors a wide array of invertebrates.

Bull Kelp
Nereocystis luetkeana

Other Names: Bullwhip kelp, ribbon kelp.
Description: Up to 20 brown blades attached to a single float, which keeps the long stipe (stalk) afloat. The stipe is attached to rock by a sturdy holdfast.
Size: Stipe to 82' (25 m) long, occasionally to 118' (36 m) long; float to 6 3/4" (17 cm) in diameter; blades to 15' (4.5 m) long, to 6" (15 cm) wide.
Habitat: On rocks, upper subtidal zone and lower; also commonly found washed up along beaches after storms.
Range: Aleutian Islands, Alaska, to San Louis Obispo County, California.
Notes: Bull kelp is one of the largest kelps in the world. Studies in Washington have shown that this species grows 5 1/2" (14 cm) per day. Historically, this kelp has been utilized by various peoples, including the Tlingits of Alaska, to make fishing line. The stipe is often used to make pickles. Bull kelp has also been used in the production of dolls and ornamental musical instruments!

SEAWEEDS

SMALL PERENNIAL KELP
Macrocystis integrifolia

OTHER NAME: Giant kelp.
DESCRIPTION: Well branched. Numerous narrow leaf-like blades, each with a small air bladder or float along the entire length. **Holdfast is a flattened** rhizoid (enlarged root-like stem) and somewhat enlarged. As it grows, it molds itself closely to the rock surface.

Holdfast and thalli (plant body).

SIZE: To 98' (30 m) long.
HABITAT: In rocky areas, very low intertidal zone to water 33' (10 m) deep.
RANGE: Kodiak Island, Alaska, to Monterey Peninsula, California; Peru; Chile.
NOTES: Small perennial kelp lives for several years in sheltered locations, and probably only 1–2 years in exposed sites. It is an important alga for several herbivores, including red urchins (see p. 203). This seaweed is very similar to giant perennial kelp (see below), which grows much larger with a pyramidal holdfast.

GIANT PERENNIAL KELP
Macrocystis pyrifera

OTHER NAME: Giant kelp.
DESCRIPTION: Brownish blades resembling those of the small perennial kelp. **Holdfast tall and pyramid shaped** (not a creeping rhizome-like stem).
SIZE: To 150' (45.7 m) long.
HABITAT: On rocks or in coarse sand from the low intertidal zone to water 265' (80 m) deep.
RANGE: Alaska to Bahía Magdalena, Baja California Sur, México.
NOTES: The giant perennial kelp forms massive kelp beds or offshore "forests of the sea," which are harvested commercially for algin, a hydrophylic colloid. Algin is an amazing substance that is used as an emulsifying, stabilizing and suspending agent in a wide range of commercial products including ice cream, chocolate milk, icings, salad dressings, toothpaste, film emulsions, paint, insecticides and oil well drilling mud.

Pacific Rockweed
Fucus gardneri

Other Names: Rock weed, bladder wrack, popping wrack, common brown rockweed; formerly *F. distichus, F. furcatus, F. evanescens*.
Description: Color varies from olive green to yellowish green. **Conspicuous midrib**. Flat blades branch regularly and dichotomously along plant. A single stipe normally originates from each holdfast.
Size: To 20" (50 cm) high; blades to $5/8$" (1.5 cm) wide.
Habitat: Attached to rocks, mid- to low intertidal zone.
Range: Aleutian Islands, Alaska, to central California.
Notes: This comnmon species can withstand both the freezing of winter and the desiccation of summer. Its swollen yellow tips or receptacles contain the gametes for reproduction. When the tide goes out, the receptacles shrink, squeezing out the gametes. When the tide returns, sperm cells are able to find and fertilize the eggs. In protected areas, this seaweed can live for 5 years. Little rockweed (see below) is a closely related species, but it has no midrib.

Little Rockweed
Pelvetiopsis limitata

Other Names: Rockweed, dwarf rockweed.
Description: Generally light green. **Flattened branches** divide evenly, similar to rockweed (see above), but there is **no midrib**.
Size: Normally to 3" (8 cm); occasionally to 7" (18 cm) tall.
Habitat: On exposed rocky areas, upper intertidal zone.
Range: Vancouver Island, BC, to Cambria, San Luis Obispo County, California.
Notes: This seaweed grows at a higher intertidal level than Pacific rockweed (see above) and so is only submerged at the highest of tides and usually on the tops of the rocks. It is therefore well adapted to spending long periods out of water.

SEAWEEDS

SPINDLE-SHAPED ROCKWEED
Pelvetia fastigiata

OTHER NAME: Formerly *Fucus fastigiatus*.
DESCRIPTION: Color varies from greenish olive to yellowish brown. Mature plants have inflated tips with a spindle-like shape. Several branches originate from each holdfast with many equal divisions. **Branches are thin** and nearly cylindrical to slightly flattened near the tips. **No midrib** is present.
SIZE: Normally to 16" (40 cm), occasionally to 36" (90 cm) tall.
HABITAT: On somewhat protected rocks from the high to mid-intertidal zone.
RANGE: Horswell Channel, BC, to Punta Baja, Baja California, México.
NOTES: This alga is often locally abundant. The size of a mature plant is a good indicator in identification. Small individuals lack the swollen tips present in little rockweed (see p. 235), a smaller species. Swollen branch tips form when the individual reaches maturity. Here reproductive bodies are released from tiny pores. Spindle-shaped rockweed can cover rocks with a thick growth drooping over the rock surface. In southern California it is the only olive green species that lacks a central rib with swollen branch tips found on high intertidal rocks.

GIANT BLADDER CHAIN
Cystoseira osmundacea

DESCRIPTION: Color varies from tan to blackish brown. Plant consists of 2 different sections: the lower branches are flattened and somewhat fern-like, while those higher up are rounded with small floats or bladders forming continuous chains.
SIZE: To 26' (8 m) tall.
HABITAT: From the low intertidal zone to water 33' (10 m) deep.
RANGE: Seaside, Oregon, to Ensenada, Baja California, México.
NOTES: The giant bladder chain is a large brown alga that can sometimes be found in tidepools in rocky habitats. This seaweed is a preferred habitat for spiral tube worms (see p. 53), which build calcareous tubes.

SARGASSUM
Sargassum muticum

OTHER NAMES: Japweed, wireweed.
DESCRIPTION: Yellowish brown stipe (stalk) with many branches, small elongated blades and **single, round floats attached individually to stipe**.
SIZE: To 6 1/2' (2 m) long.
HABITAT: In protected rocky areas, low intertidal zone to subtidal water 16' (5 m) deep.
RANGE: Entire Pacific coast.
NOTES: Sargassum was accidentally introduced from Japan along with oysters in the 1930s and has spread all along the Pacific coastline. It is often found washed up on the beach.

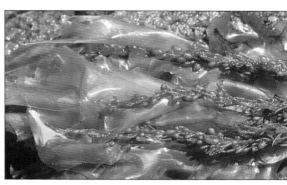
Sargassum and sea lettuce.

RED ALGAE
Phylum Rhodophyta

Red algae contain chlorophyll, but red and blue pigments are also present, giving these algae a reddish tinge. The life cycle of red algae usually involves three stages, rather than two as in other algae.

RED FRINGE
Smithora naiadum

DESCRIPTION: Purplish red to deep purple paper-thin blades.
SIZE: 3/4–4" (2–10 cm) wide.
HABITAT: On leaves of surf-grasses and eel-grasses (p. 253), in intertidal zone.
RANGE: Alaska to México.
NOTES: Red fringe is a distinctive seaweed that is only one cell thick. It is attached to eel-grass or surf-grass by a short, narrow portion of the blade. This delicate seaweed is found most often in late summer and fall, when it is abundant in many tidepools.

SEAWEEDS

LAVER
Porphyra sp.

OTHER NAMES: Purple laver, red laver, dulse, nori, wild nori.
DESCRIPTION: Color varies from purple to green. Single broad blade, often irregular in shape, has ruffled margins. Often found in dense clusters.
SIZE: 8–60" (20–150 cm) long.
HABITAT: In rocky areas, intertidal to upper subtidal zones. Can be epiphytic (attached to an organism with no harm to that organism) within these zones.
RANGE: Alaska to México.
NOTES: Laver is rubbery and gelatinous in texture, once out of the water. It is edible, and most species are considered to be very tasty as well as having large amounts of vitamins A and C. In Japan, this seaweed is called nori and has been collected and used as food for 1,000 years. Today laver is big business in Japan, worth $1 billion annually.

RUBBER THREADS
Nemalion helminthoides

OTHER NAME: Formerly *Fucus elminthoides*.
DESCRIPTION: Golden brown to reddish brown in color with a smooth, soft texture. The thalli (plant body) is generally unbranched but some branching may be present. Several plants often grow clustered together.
SIZE: Usually 8–18" (20–45 cm) tall and $1/32$–$1/8$" (1–4 mm) in diameter.
HABITAT: On sheltered rocks in the high to mid-intertidal zone.
RANGE: Aleutian Islands, Alaska, to Baja California, México.
NOTES: The gametophytic (gamete producing) stage of this alga bears a close resemblance to an earthworm. This species is often found in clusters in the springtime. The sporophytic (spore producing) stage is visible only under the microscope. Rubber threads also occurs in the north Atlantic, Japan, Australia and New Zealand.

Hairy Seaweed
Cumagloia andersonii

OTHER NAME: Formerly *Nemalion andersonii*.
DESCRIPTION: Color varies from brown to purple. Tough, elongated cords, with numerous fibers along the length, hang from rock substrate at low tide.
SIZE: To 3' (90 cm) tall.
HABITAT: On rocks, high intertidal zone.
RANGE: Hope Island, BC, to Cabo Colnett, Baja California, México.
NOTES: This annual species grows back on the same rocks year after year. It reaches its maximum size in August.

Bleached Brunette
Cryptosiphonia woodii

OTHER NAMES: Formerly *C. grayana*, *Pikea woodii*.
DESCRIPTION: Color varies from olive brown to blackish purple; golden when branches are fertile, late in the season. Plant is coarse overall, with a slender axis and much branching.
SIZE: To 10" (25 cm) tall, occasionally to 14" (35 cm).
HABITAT: On rocks in the mid-intertidal zone.
RANGE: Unalaska Island, Alaska, to San Pedro, Los Angeles County, California.
NOTES: Bleached brunette is often one of the first seaweeds to colonize new areas. It grows in clusters and is commonly found in both exposed and protected sites.

SEAWEEDS

ENCRUSTING CORALLINE ALGAE

Lithothamnion sp.

OTHER NAMES: Encrusting coral, rock crust; formerly *Lithothamnium* sp.
DESCRIPTION: Light rose in color. Thin, crust-like shapes resemble lichens.
SIZE: To 1/16" (2 mm) thick, and can cover large areas.
HABITAT: In rocky areas, low intertidal zone to subtidal waters.
RANGE: Aleutian Islands, Alaska, to La Jolla, San Diego County, California.
NOTES: Rocks in tidepools and the shells of various gastropods are often covered with this alga. Several invertebrates feed on it, including whitecap limpet (see p. 75) and lined chiton (p. 58).

STALKED CORALLINE DISK

Clathromorphum parcum

OTHER NAMES: Formerly *Polyporolithon parcum, Lithothamnium parcum*.
DESCRIPTION: Color varies from lavender to purplish. The **disc-shaped body** is smooth and slightly convex. Species is attached to a host plant by a **central stalk**.
SIZE: To 3/4" (2 cm) in diameter.
HABITAT: Intertidal zone to shallow subtidal waters.
RANGE: Queen Charlotte Islands, BC, to San Luis Obispo County, California.
NOTES: This alga grows on a wide variety of articulated coralline algae, including tidepool coralline alga (see p. 241). It grows as single plants that do not crowd together.

Stone Hair
Lithothrix aspergillum

DESCRIPTION: Overall color is lavender-pink. Articulated coralline alga with elongated branches that are often densely branched into tufts stemming from a crustose base.
SIZE: To 5" (13 cm) tall and a mere 1/32" (less than 1 mm) in diameter.
HABITAT: In sandy areas on rocks, mid-intertidal tidepools to waters 43' (13 m) deep.
RANGE: Cook Inlet, Alaska, to Isla Magdalena, Baja California, México.
NOTES: Stone hair is a delicate looking seaweed with fine fronds. Its individual segments are short and minute, making it easy to identify. This elegant species attaches to both rocks and animals.

Tidepool Coralline Alga
Corallina officinalis var. *chilensis*

OTHER NAMES: Common coral seaweed, tide pool coral, tall coralline alga; formerly *Corallina chilensis*.
DESCRIPTION: A white to pink or purple **calcareous** seaweed with jointed segments and many flattened, pinnate or **feather-like branches** with branching of equal size. **Segments on the central axis** are slightly flattened and **elongated**, to $1/16$–$1/8$" (1.5–3 mm) in length.
SIZE: To 6" (15 cm) tall.
HABITAT: In tidepools or low intertidal zone to water 33' (10 m) deep.
RANGE: Prince William Sound, Alaska, to Chile.
NOTES: This common articulated seaweed can often be seen in tidepools. Until the early 18th century it was used as a vermifuge (a substance that expels parasites from the body). An encrusting base adheres tightly to rock, from which the branches grow. Remarkably, this and several similar algae contain a very high percentage of calcium carbonate, the same material that makes up clam and snail shells.

SEAWEEDS

COARSE BEAD-CORAL ALGA
Calliarthron tuberculosum

OTHER NAMES: Formerly *C. regenerans, C. cheilosporioides, C. setchelliae, Corallina tuberculosa*.
DESCRIPTION: A pink to purple calcareous seaweed with **jointed segments** that may have **both cylindrical and flattened shapes**. Branching is irregular. The **conceptuals** (small reproductive cavities that may have surface bumps) occur **on the edges of the segments** and occasionally the flat portions. Branches grow from the main stem at more than a 60-degree angle. Beads or segments are generally coarse.
SIZE: To 8" (20 cm) tall.
HABITAT: From tidepools in the low intertidal zone to subtidal waters.
RANGE: Alaska to Isla Cedros, Baja California, México.
NOTES: Beachcombers often find this beautifully jointed alga in tidepools. It is the most common erect coralline algae along California's coastline.

NAIL BRUSH SEAWEED
Endocladia muricata

OTHER NAME: Sea moss.
DESCRIPTION: Color ranges from dark red to black or greenish brown. Long filaments, covered with tiny spines, grow in tufts.
SIZE: To $1^{1/2}$–3" (4–8 cm) tall.
HABITAT: On rocks, high intertidal zone.
RANGE: Alaska to Punta Santo Tomás, Baja California, México.
NOTES: This alga's name describes it very well. It is a hardy species, able to withstand long periods out of water and locally common on the tops and sides of larger rocks in many areas.

BROAD IODINE SEAWEED
Prionitis lyallii

OTHER NAMES: Iodine seaweed, Lyall's iodine seaweed.
DESCRIPTION: Color ranges from dull brown to reddish purple. Small root-like holdfast holds the flattened blade and bladelets attached along the margin.
SIZE: To 14" (35 cm) long.
HABITAT: On rocks in tidepools, mid-intertidal zone and deeper into upper subtidal zone. Coarse sand often covers the rock this species attaches to.
RANGE: Southern BC to Punta María, Baja California, México.
NOTES: This seaweed smells like bleach. It is often found in sandy areas.

RED SEA-LEAF
Erythrophyllum delesserioides

DESCRIPTION: Blades are dark red drying to black with a midrib extending through the entire blade. Lateral veins arise from the midrib in an alternating pattern.
SIZE: Normally to 20" (50 cm) tall but can extend to 40" (1 m).
HABITAT: On rocks in the low intertidal zone to shallow subtidal depths.
RANGE: Kayak Island, eastern Gulf of Alaska, to Shell Beach, San Luis Obispo County, California.
NOTES: Red sea-leaf is found on exposed coasts and in situations such as surge channels where rough water predominates. Its blades are torn away more and more as the summer progresses until only the midrib and side veins remain. Late in the season, short nipple-like papillae, which are used for reproduction, cover most of this plant.

SEAWEEDS

SUCCULENT SEAWEED
Sarcodiotheca gaudichaudii

OTHER NAMES: Formerly *Agardhiella tenera*, *Neoagardhiella baileyi*, *Neoagardhiella gaudichaudii*.
DESCRIPTION: Color varies from pink to red. **Slender, somewhat fleshy branches** and many branchlets tapering to sharp points.
SIZE: To 16" (40 cm) long.
HABITAT: In rocky areas, low intertidal zone to subtidal waters 60' (18 m) deep, and sometimes in tidepools.
RANGE: Alaska to Chile.
NOTES: This common species often grows in clumps.

PEPPERED SPAGHETTI
Gracilaria pacifica

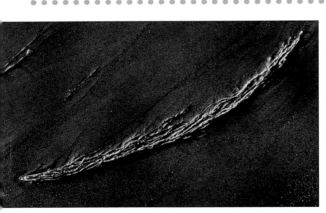

OTHER NAMES: California limu, sewing thread; formerly *G. confervoides*, *G. verrucosa* by some authors.
DESCRIPTION: The yellowish to reddish brown or black **branches are slender** and spaghetti-like, often **with small dark bump-like reproductive structures** attached.
SIZE: To 20" (50 cm) tall.
HABITAT: On rocks in protected sites from the mid-intertidal zone to shallow subtidal waters that may be partially covered in sand.
RANGE: Prince William Sound, Alaska, to Baja California, México.
NOTES: Peppered spaghetti is an edible seaweed that has also been commercially harvested as a source of agar, a substance used as a thickening agent. This distinctive species is sometimes found stranded on the shore in large numbers. It is found along the Atlantic coast and in other locations throughout the world.

Flat-Tipped Wire Alga

Mastocarpus jardinii

Other Names: Formerly *Gigartina agardhii* (gametophyte phase); *Petrocelis franciscana* (sporophyte phase).
Description: Gametophyte/sexual stage: blade is reddish brown. Small clumps of narrow and **wiry blades with flattened ends** and several even divisions. Groups of nipple-like papillae are present on the blade surface.
Size: Gametophyte (blade) stage to 6" (15 cm) tall.
Habitat: From the high to mid-intertidal zone.
Range: Northern BC to San Luis Obispo County, California.
Notes: Like several other seaweeds, the flat-tipped wire alga displays distinct stages in its life history. The gametophyte (blade) stage is the most obvious while the sporophyte (asexual) stage is crust-like in nature. This sporophyte stage is often referred to as the *Petrocelis* stage since it was originally identified as a separate species, *Petrocelis franciscana*.

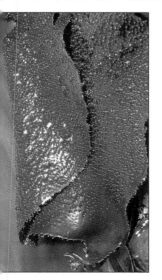

Turkish Towel

Chondracanthus corymbiferus

Other Names: Formerly *Gigartina corymbifera*, *Mastocarpus corymbiferus*.
Description: Brick red to purplish red blades, often iridescent when underwater. Blade surface is covered with tiny rasp-like projections.
Size: To 20" (50 cm) long, 8" (20 cm) wide.
Habitat: Lower intertidal zone to water 65' (20 m) deep.
Range: Barkley Sound, Vancouver Island, BC, to Cabo San Quíntin, Baja California, México.
Notes: This species is a source of carrageenan, a stabilizer for a wide range of products including cottage cheese and printer's ink, and has been experimentally farmed to produce this substance.

SEAWEEDS

PAPILLATE SEAWEED/ SEA TAR

Mastocarpus papillatus

OTHER NAMES: Crust stage: Sea tar, tar spot, sea film; formerly *Petrocelis franciscana, Petrocelis middendorffii*. **Blade stage:** Crisp leather, Turkish washcloth; formerly *Gigartina papillata, Gigartina cristata*.

DESCRIPTION: Crust stage: Dark red-brown to black smooth crust grows on rock. **Blade stage:** Upright, smooth blade ranging in color from yellow-brown to dark purple or even black. Blades are palm-like in shape, normally with several branches. Flat, irregularly branched blades later develop small growths or projections.

SIZE: Crust stage: to 3.3' (1 m) in diameter. **Blade stage:** to 6" (15 cm) tall.

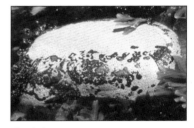

Crust stage.

HABITAT: In rocky areas, high intertidal (near rockweed) to low intertidal zone.
RANGE: Aleutian Islands, Alaska, to Punta Baja, Baja California, México.
NOTES: This highly variable species had scientists mystified for years. It was originally thought that the 2 distinct stages were separate species. Only recently was it discovered that both stages were part of the same life cycle. The crust stage reproduces asexually to produce spores, which develop into the blade stage. Blades are either male or female and reproduce sexually. Studies have shown that the crustose stage of this alga grows at an extremely slow rate, about $1/2$" (1.3 cm) per year, and that larger specimens could reach the incredible age of 90 years old.

Iridescent Seaweed

Mazzaella splendens

OTHER NAMES: Rainbow seaweed; formerly *Iridophycus splendens, Iridaea cordata*.
DESCRIPTION: Blades vary in color from green to bluish purple. Blades often have a split or lobed shape, and appear iridescent in sunlight.
SIZE: Blades to 12" (30 cm) long, occasionally to 3' (1 m).
HABITAT: In rocky, exposed areas, low intertidal zone to water 23' (7 m) deep.
RANGE: Northern southeast Alaska to Baja California, México.
NOTES: This striking species is memorable for its dazzling iridescence, which is especially vivid when touched with sunlight underwater. Various snails feed on it, which is why several holes are sometimes found on the blades.

Sea Sacs

Halosaccion glandiforme

OTHER NAMES: Dead man's fingers, sea sacks, salt sacs, sea nipples.
DESCRIPTION: Color varies considerably from yellowish brown to olive brown and reddish purple. Long, hollow sacs are attached in clusters.
SIZE: To 12" (30 cm) long, sacs to $1^{1/2}$" (4 cm) in diameter.
HABITAT: On both exposed and protected rocky areas, mid-intertidal zone.
RANGE: Aleutian Islands, Alaska, to Point Conception, California.
NOTES: Young sea sacs are normally purple, and slowly change to light yellow as they grow older. Seawater normally fills or nearly fills the sacs when they are submerged. A portion of this water is released through tiny pores when the tide recedes.

SEAWEEDS

DELICATE SEA LACE
Microcladia coulteri

DESCRIPTION: Deep red to rose in color. Appears delicate and lacy overall and grows in one plane. 5 to 7 orders of branching, in alternate fashion.
SIZE: To 16" (40 cm) tall.
HABITAT: Mid-intertidal zone to water 33' (10 m) deep, and often epiphytic (living on another species without harming it).
RANGE: Aleutian Islands, Alaska, to San Diego, California.
NOTES: This beautiful seaweed is common locally. It is sometimes found growing on a variety of red and brown seaweeds, including feather boa kelp (see p. 231) and broad iodine seaweed (p. 243). Another very similar species, California sea lace *Microcladia californica*, is found only in California waters from San Francisco to Monterey Peninsula and possibly San Diego. These two species cannot be identified with certainty in the field; identification must be made in a lab.

HALF-FRINGED SEA FERN
Neoptilota densa

OTHER NAME: Formerly *Ptilota densa*.
DESCRIPTION: Color is dark red overall. Branches are fern-like, densely fringed with irregular branching. **Branchlets are toothed along one edge only**.
SIZE: To 12" (30 cm) tall.
HABITAT: Low intertidal zone to 50' (15 m). **Usually epiphytic on coralline algae** (attached to the algae without harming them).
RANGE: Tomales Bay, Marin County, California, to Bahía Rosario, Baja California, México.
NOTES: Half-fringed sea fern is normally found growing on various articulated coralline algae such as coarse bead-coral alga (see p. 242). Its beautiful, delicate branches are fern-like in shape, hence its common name.

Winged Fronds
Delesseria decipiens

OTHER NAMES: Baron delessert, winged rib.
DESCRIPTION: Color is normally bright red but varies to yellowish. The main stem displays a prominent midrib while **wing-like extensions** are found **on the branches and leaflets**. Each branch arises from the midrib of a larger branch. However, the wings and many branches of specimens found on the beach are often eroded away. A minute disk holds this alga in place.
SIZE: To 14" (35 cm) tall.
HABITAT: From the low intertidal zone to water 60' (18 m) deep.
RANGE: Prince William Sound, Alaska, to San Luis Obispo County, California.
NOTES: This splendid seaweed has been found to photosynthesize at its maximum when in indirect sunlight, while harsh direct sunlight may in fact damage it. Winged fronds contain red water-soluble pigments called phycobilins, which are uncommon in algae. The elegant movement of the delicate blades drifting in quiet waters is truly beautiful to observe.

• •

Fringed Hidden Rib
Cryptopleura ruprechtiana

OTHER NAMES: Hidden rib, ruche; formerly *Nitophyllum ruprechtianum, Botryoglossum farlowianum, B. ruprechtianum*.
DESCRIPTION: Color varies from rose red to purplish red. Its **blades are flat and deeply divided** into antler-like shapes. A **midrib** is present at the base but **becomes hidden** as it gives way to a network of veins in each blade. The **edge** of the blade often **displays a fringe** of tiny semicircular shapes.
SIZE: To 20" (50 cm) tall.
HABITAT: In exposed areas from the low intertidal zone to 100' (30 m).
RANGE: Northern southeast Alaska to Baja California, México.
NOTES: This distinctive alga is common and widespread. It forms large patches in the low intertidal zone. The fringed hidden rib holds male and female gametes as separate individuals.

SEAWEEDS

BLACK TASSEL
Pterosiphonia bipinnata

OTHER NAMES: Formerly *Polysiphonia bipinnata, Pterosiphonia robusta*.
DESCRIPTION: Color varies from bright red to brownish red. **Branches are very fine, cylindrical and distichous** (arranged in 2 rows on opposite sides of the axis). The branches at the ends of the fronds grow in a single plane.
SIZE: Normally to 5" (12 cm) tall, but can reach 10" (25 cm).
HABITAT: Mid-intertidal zone to upper subtidal waters.
RANGE: Bering Sea, Alaska, to San Pedro, Los Angeles County, California.
NOTES: Black tassel occurs in both exposed and sheltered sites in the spring. It is often found dried out in the sun for long periods, which causes its outer layers to turn light pink. The species also grows as an epiphyte on other seaweeds (attached to them without harming them).

BLACK PINE
Neorhodomela larix

OTHER NAME: Formerly *Rhodomela larix*.
DESCRIPTION: Distinctive brownish black color. Round branches and branchlets in a uniform cylindrical shape. Clusters often grow from the same attachment point.
SIZE: To 12" (30 cm) long.
HABITAT: In rocky areas, low intertidal zone.
RANGE: Aleutian Islands, Alaska, to Baja California, México.
NOTES: This species thrives on rock in sandy, exposed situations such as the outer coast and in exposed rocky tidepools. Here it can grow into mats of many individuals. It is also more tolerant of sediment than other seaweeds.

FLOWERING PLANTS
Phylum Anthophyta

Common Pickleweed
Salicornia virginica.

FLOWERING PLANTS

The flowering plants found in the intertidal zone lack the bright colors of flowering plants elsewhere, but they provide an important habitat for a wide array of both invertebrates and vertebrates (fish).

COMMON PICKLEWEED
Salicornia virginica

OTHER NAMES: American glasswort, sea asparagus, perennial saltwort, Pacific samphire; *S. pacifica*.

DESCRIPTION: A green, **fleshy succulent with jointed stems** and leaves reduced to minute scales. A **perennial** with flowers extending to the tips of the branches.

SIZE: To 12" (30 cm) tall and stems that may trail to 39" (1 m).

HABITAT: On sheltered shores at the high tide mark and especially common in saltwater marshes and tide flats.

RANGE: Alaska to México; Atlantic coast.

NOTES: This common, widespread plant is one of several pickleweeds that occur in California. It is accompanied by European pickleweed *S. europaea*, a smaller, reddish, bushy annual that grows in crowded conditions in northern California. Southern pickleweed *S. bigelovii*, a tall succulent that branches at its base, is another annual that is very common in southern California. Pickleweeds have been used as food in cultures throughout the world for many generations. They are often pickled (hence their name), as well as being eaten as a fresh (and salty) veggie.

SURF-GRASS

Phyllospadix sp.

DESCRIPTION: Long, wiry blades with oval or round cross-section.
SIZE: Leaves normally to 5' (1.5 m) long, blade to 1/8" (4 mm) wide.
HABITAT: On rock in exposed areas, low intertidal to upper subtidal zones.
RANGE: Alaska to México.
NOTES: This common plant is found in just about all open coast sites. It provides food and shelter to many other plant and animal species. The sweet rhizomes (root-like stems) of both surf-grass and eel-grass (see below) were traditionally eaten fresh or dried for the winter by various Native cultures.

EEL-GRASS

Zostera sp.

DESCRIPTION: Long, flat blades. Roots grow in mud.
SIZE: Leaves normally to 8' (2.5 m) long, 1/2" (1.2 cm) wide.
HABITAT: In quiet bays with mud bottom, low intertidal zone to water 100' (30 m) deep.
RANGE: Alaska to México.
NOTES: Like surf-grass (above), this is a species on which many other plants and animals depend for food and shelter. People have harvested the seeds of this species as they harvest wheat.

BEST BEACHCOMBING SITES IN CALIFORNIA

The California coastline has hundreds of seashores for the beachcomber to explore, with extraordinary landscapes that include islands, sea arches, coves, estuaries, tidepools, sand and mud beaches. A few excellent viewing sites are described in this chapter.

These sites are protected areas that require the respect of visitors and a determined public effort to keep them healthy for future generations to experience. They are wild natural areas whose ecosystems are fragile, no matter how rugged the landscape appears. Look out for your own safety and for the safety of the living things you see on the seashore.

A California sport fishing license allows you to take some game fish, crabs, abalone and clams, but you may take no other organisms unless you have a scientific collecting permit.

Key to Site Descriptions

Exposed Subjected to the direct action of waves

Semi-exposed Somewhat sheltered from large waves

Protected Not subjected to the full force of wave action

Sandy Mostly sand deposited on a beach

Mud Primarily a mud base, in sheltered areas, with a low volume of sand

Sand/Mud Mixture of sand and mud

Gravel Accumulation of small rocks or pebbles

Rock Mostly bedrock, rocky outcrops, boulders or rocks

Tidepools Having one or more pools of salt water, varying in size, once the tide goes out

Opposite: Salt Point State Park, Sonoma County.

NORTH COAST

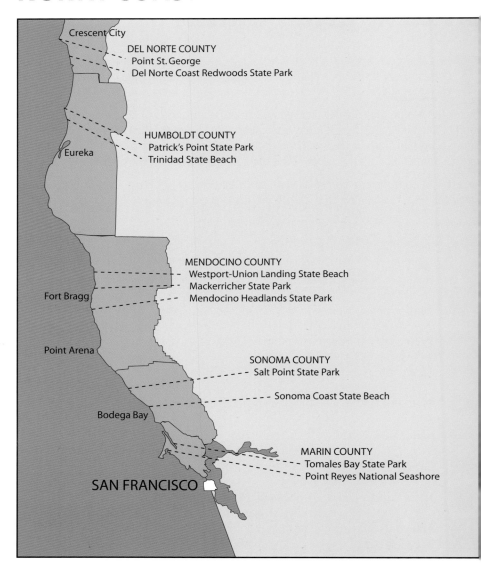

DEL NORTE COUNTY

   Point St. George

SITE DESCRIPTION: Exposed rocky area with tidepools.
NEAREST CITY: Crescent City.
ACCESS: From Crescent City, travel North on Highway 101, take West Washington Boulevard and follow it to the parking area for this site. A short walk takes you to the point and nearby tidepools.
NOTES: You can enjoy the seashore and see a variety of rock formations here. Watch for sandstone in which various clam shells are fossilized into the rock, evidence of the marine life in the area thousands of years ago.

A wide variety of marine life is found at Point St. George, including purple shore crab (see p. 186), giant green anemone (p. 30), aggregating anemone (p. 29), giant rock scallop (p. 131), hairy hermit (p. 173) and purple star (p. 197). Recent empty shells to be found here include the giant western nassa (p. 104), Japanese littleneck (p. 149) and Pacific gaper (p. 138), and the empty exoskeleton of the dungeness crab (p. 183).
FOR MORE INFORMATION: Write to Del Norte County Parks and Recreation Department, 840 9th Street, Crescent City, CA. Tel: (707) 464-7230.

  Del Norte Coast Redwoods State Park

SITE DESCRIPTION: Semi-exposed site with scattered rocks, boulders and tidepools.
NEAREST CITY: Crescent City.
ACCESS: Take Highway 101 south from Crescent City, 7 miles to Wilson Beach. Watch for the sign at the south end of the park.
NOTES: Wilson Beach and Creek are situated in the often fog-enveloped Del Norte Coast Redwoods State Park. Turn off at the picnic area with a small sandy beach, where you can stop for a rest or picnic after traveling through the world-renowned Redwood National Park

complex. Del Norte Coast Redwoods State Park is one part of this World Heritage Site & Biosphere Preserve.

At the tidepools at Wilson Beach, or False Klamath Cove, you can observe the purple star (see p. 197), sunflower star (p. 197), leather star (p. 194), black turban (p. 81) and red rock crab (p. 182) among the rocks. Exercise caution—various algae cover many of the rocks here, making them very slippery. Bread crumb sponge (p. 18) also lines the shaded side of some rocks. The state campground is only 2 miles to the east.

FOR MORE INFORMATION: Write to: Del Norte Coast Redwoods State Park, c/o Redwood Coast Satellite Sector Office, 1375 Elk Valley Road, Crescent City, CA 95531. Tel: (707) 464-6101 ext 5120. Web site: http://cal-parks.ca.gov/

HUMBOLDT COUNTY

Patrick's Point State Park

SITE DESCRIPTION: Exposed area with both sandy and rocky sites.
NEAREST CITY: Trinidad.
ACCESS: From Trinidad, drive 5 miles north on Highway 101 and take the Patrick's Point exit.
NOTES: Patrick's Point State Park features the spectacular Agate Beach, a sandy beach favored for picnicking, tidepooling and beachcombing for agates. Palmer's Point, a second intertidal site, features a rocky area with great tidepools. From either parking area you can take a short trail to these intertidal areas.

A few of the interesting invertebrates found here are beautiful pink-colored moonglow anemone (see p. 28), aggregating anemone (p. 29), painted anemone (p. 33), red nudibranch (p. 117), opalescent nudibranch (p. 124), giant western nassa (p. 104), grainyhand hermit (p. 172) and leather star (p. 194).

Be sure to stop in at the visitor center, located near the entrance, with its various displays and natural history books for sale. This park is also a favorite for camping and picnicking.

FOR MORE INFORMATION: Write to: Patrick's Point State Park, 4150 Patrick's Point Drive, Trinidad, CA 95570. Tel: (707) 677-3570. Web site: http://cal-parks.ca.gov/

North Coast

  ## Trinidad State Beach

SITE DESCRIPTION: Protected site with boulders and tidepools.
NEAREST CITY: Trinidad.
ACCESS: Exit Highway 101 at Trinidad and drive through town for about 1/4 mile. Turn north on Stagecoach Road to reach the parking area.
NOTES: Trinidad State Beach is often shrouded in fog. This surreal landscape features sea stacks and an island that is accessible at low tide. From the parking lot, take the winding trail to the beach, and walk north a short distance to reach the intertidal area. Here a rich variety of seaweeds abound, making the rocks slippery. The tidepools may hold frosted nudibranch (see p. 121), leather star (p. 194), smooth bay shrimp (p. 170) and giant green anemone (p. 30), while purple stars (p. 197) cling to the rocks. Hairy hermits (p. 173) can be spotted higher up on the shoreline, and dungeness crab (p. 183) and eccentric sand dollars (p. 204) may be seen in the sandy areas.

While in Trinidad, be sure to stop in at the small aquarium maintained by the Humboldt State University Marine Lab. Its excellent free displays, aquariums and touch tanks are open to the public.

FOR MORE INFORMATION: Write to: Trinidad State Beach, 4150 Patrick's Point Drive, Trinidad, CA 95570. Tel: (707) 677-3570. Web site: http://cal-parks.ca.gov/

MENDOCINO COUNTY

  ## Westport-Union Landing State Beach

SITE DESCRIPTION: Exposed with boulders, rocks and tidepools.
NEAREST CITY: Westport.
ACCESS: From Westport, drive north 2 miles on Highway 1.
NOTES: This state beach features 2 miles of beach and is divided into two sections along the highway. The tidepools are accessible from the southern section at Abalone Point. As the name suggests, this is a great place to see

259

abalone! Look for them under overhangs and in similar situations at the lowest of tides. Abalone enthusiasts often gather at this site when the season is open. Check for current regulations with the California Fish and Game Department.

Intertidal life abounds here, with opportunities to see giant green anemone (p. 30), proliferating anemone (p. 31), painted anemone (p. 33), moonglow anemone (p. 28), orange-spotted nudibranch (p. 115), Monterey dorid (p. 118), opalescent nudibranch (p. 124), leather star (p. 194) and lightbulb tunicate (p. 210). Seaweeds, including bull kelp (p. 233), are also abundant here.

The camping area is situated on a bluff with fabulous ocean views.

FOR MORE INFORMATION: Write to: Westport-Union Landing State Beach, c/o Russian River-Mendocino Management District, 25381 Steelhead Boulevard, P.O. Box 123, Duncans Mills, CA 95430. Tel: (707) 865-2391.
Web site: http://cal-parks.ca.gov/

Mackerricher State Park

SITE DESCRIPTION: Exposed site with a rocky edge, large boulders and tidepools.
NEAREST CITY: Fort Bragg.
ACCESS: From Fort Bragg, drive 3 miles north on Highway 1 and turn west at Cleone.
NOTES: Visitors to this popular park enjoy a wide range of activities, including camping, fishing and tidepooling. To reach the tidepool area, drive through the park and follow the signs to the day use and beach area. Walk a short distance along the trail and boardwalk to the seal-watching station, and take the stairs down to the tidepool area. Here you can take advantage of guided activities conducted by park rangers. While tidepooling, be sure not to disturb the harbor seals lying on the rocks or basking in the sun.

Marine life to be seen at this site includes orange cup coral (p. 28), painted anemone (p. 33), proliferating anemone (p. 31), giant Pacific chiton (p. 69), white spotted sea goddess (p. 119), opalescent nudibranch (p. 124), spot-bellied rock crab (p. 183), leather star (p. 194), bat star (p. 193), six-rayed star (p. 195), sunflower star (p. 197) and red sea urchin (p. 203). The sea palm (p. 232) is also found here, indicating that this area is exposed to the pounding surf.

FOR MORE INFORMATION: Write to: Mackerricher State Park, c/o Russian River-Mendocino Management District, 25381 Steelhead Boulevard, P.O. Box 123, Duncans Mills, CA 95430. Tel: (707) 865-2391. Web site: http://cal-parks.ca.gov/

Mendocino Headlands State Park

SITE DESCRIPTION: Exposed area with rocks, sandy beach and tidepools.
NEAREST CITY: Mendocino.
ACCESS: The park surrounds the town of Mendocino, just off Highway 1.
NOTES: Mendocino Headlands State Park offers spectacular views of a rugged coastline that is rich in islands, sea arches, coves and tidepools. A staircase at the south end leads to a small sandy beach—a great spot for sunbathers. For tidepooling, carefully follow one of the trails that lead down to the tidepools.

Marine life at this site includes the colorful red-beaded anemone (p. 32), painted anemone (p. 33), red abalone (p. 72), sunflower star (p. 197), leather star (p. 194), Pacific blood star (p. 195), bat star (p. 193), hundreds of purple sea urchins (p. 203) and a few red sea urchins (p. 203).

Be sure to stop in at the Ford Museum, a visitor center and museum located in Mendocino, where excellent displays are open to the public.

FOR MORE INFORMATION: Write to: Mendocino Headlands State Park, c/o Russian River-Mendocino Management District, 25381 Steelhead Boulevard, P.O. Box 123, Duncans Mills, CA 95430. Tel: (707) 865-2391. Web site: http://cal-parks.ca.gov/

SONOMA COUNTY

  ## Salt Point State Park

SITE DESCRIPTION: Exposed area with rocks, boulders and tidepools.
NEAREST CITY: Guerneville.
ACCESS: From Jenner, drive 20 miles north on Highway 1.
NOTES: Salt Point State Park is a 60,000-acre park that features a spectacular rocky shoreline, pigmy forest and tidepools at Crystal Cove. It is a viewing site for a variety of wildlife, especially the brown pelican and several mammals.

Hiking, camping, tidepooling and diving are popular activities at this park. As with all coastal areas, caution is recommended as part of the coastline is hazardous. And please note that all marine life is protected by law in the Crystal Cove Marine Reserve.

Intertidal creatures at this wild site include the orange cup coral (see p. 28), strawberry anemone (p. 35), giant Pacific chiton (p. 69), sunflower star (p. 197) and bat star (p. 193). Seaweeds are abundant and cover the boulders, so they are slippery. This makes for slow going, but the many and varied species present make it all worthwhile.

A visitor center is located at Crystal Cove.

FOR MORE INFORMATION: Write to: Salt Point State Park, c/o Russian River-Mendocino Management District, 25381 Steelhead Boulevard, P.O. Box 123, Duncans Mills, CA 95430. Tel: (707) 847-3221. Web site: http://cal-parks.ca.gov/

Sonoma Coast State Beach

SITE DESCRIPTION: Exposed site with rocks, boulders, and tidepools.
NEAREST CITY: Bodega Bay.
ACCESS: From Jenner, drive south on Highway 1 about 3 1/2 miles.
NOTES: Sonoma Coast State Beach stretches along the coast for several miles. When the fog rolls in, it's a sight that's not to be forgotten. Shell Beach offers excellent opportunities to observe intertidal life and is noted as a prime fishing spot. A short trail leads from the Shell Beach parking area to the intertidal site.

At this picturesque site you may see giant green anemone (see p. 30), aggregating anemone (p. 29), purple star (p. 197), purple shore crab (p. 186) and black turban (p. 81). Abundant seaweeds also cling to the rocks here.

Be aware that this site and all north coast beaches are not recommended for swimming. Strong rip currents, heavy surf and sudden swells make it dangerous even to play in the surf.

FOR MORE INFORMATION: Write to: Sonoma Coast State Beach, c/o Russian River-Mendocino Management District, 25381 Steelhead Boulevard, P.O. Box 123, Duncans Mills, CA 95430. Tel: (707) 847-3221. Web site: http://cal-parks.ca.gov/

MARIN COUNTY

Tomales Bay State Park

SITE DESCRIPTION: Protected site; sand and mud beach with rocks.
NEAREST CITY: Inverness.
ACCESS: Travel 4 miles north of Inverness on Pierce Point Road. The park is situated on the west side of the bay.
NOTES: The park road winds through a virgin stand of Bishop pines to reach Hart's Desire Beach, one of 4 beaches found in this impressive park. This is a very sheltered bay, where popular activities include canoeing and kayaking. An orange, sponge-like blue-green algae occurs here, attached to rock edges in the splash zone to the south. Walk north for the best tidepool viewing. Here shag-rug nudibranch (see p. 123) feed on aggregating anemone (p. 29), and you may see the fat innkeeper worm (p. 56), green crab (p. 184), Monterey dorid (p. 118) and Atlantic oyster drill (p. 102).
FOR MORE INFORMATION: Write to: Tomales Bay State Park, Star Route, Inverness, CA 94937. Tel: (415) 669-1140. Web site: http://cal-parks.ca.gov/

Point Reyes National Seashore

SITE DESCRIPTION: Protected sites.
NEAREST CITY: San Rafael.
ACCESS: From San Rafael, travel north on Sir Francis Drake Boulevard for about 18 miles to reach Highway 1, from which the site can be reached.
NOTES: This large, popular preserve includes seashore, estuary, grasslands, impressive cliffs, forested sections and spectacular views. Visitors have to hike to both of the two very different intertidal sites here, Drakes Estro and Palomarin Beach. Drakes Estro is an estuary, reached by hiking a well-marked 9-mile trail. Here, where this trail reaches the protected sand and mud shore, you may find pickleweed (see p. 252) and other inhabitants of this sheltered environment. An oyster farm is also

located here. Palomarin Beach has impressive tidepools at very low tides. Visitors hike 1 1/2 miles to reach this site. Species occurring here include bat star (p. 193), purple star (p. 197) and black Katy chiton (p. 70). Many other hikes of varying lengths lead to a variety of destinations at this preserve.

Please note that the waters here are dangerous. Caution should be exercised if you swim in this area.

Be sure to visit the Bear Valley Visitor Center located near Olema, with lots of information, books and displays, trail maps and more.

FOR MORE INFORMATION: Write to: Point Reyes National Seashore, Point Reyes, CA 94956-9799. Tel: (415) 663-1092 (Bear Valley Visitor Center). Web site: http://www.nps.gov/pore/

CENTRAL COAST

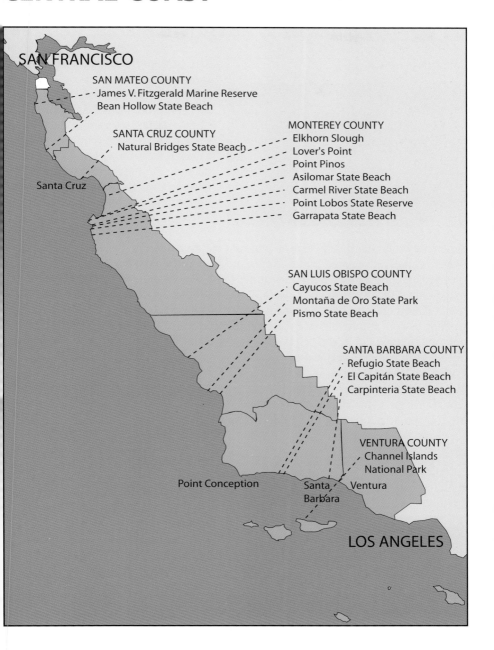

SAN MATEO COUNTY

    James V. Fitzgerald Marine Reserve

SITE DESCRIPTION: Exposed, rocky shoreline and sandy beach.
NEAREST CITY: Moss Beach.
ACCESS: From Moss Beach, take California Avenue west to the T-intersection and turn on North Lake Street to reach the parking lot. A short walk takes you to the impressive reefs.
NOTES: This reserve offers a fabulous opportunity to observe intertidal life. The reefs here are protected as part of the Monterey Bay National Marine Sanctuary, so no marine life may be removed without a permit. Scientists have conducted studies at this site for many years and have discovered more than 25 new species here. Some of the marine creatures to be seen here are red abalone (see p. 72), rough limpet (p. 78), Hopkin's rose (p. 114), spot-bellied rock crab (p. 183) and black-clawed crab (p. 184). Seaweeds are abundant here as well.

Park rangers and roving interpreters are often present to answer questions and point out marine life to visitors. Be sure to ask about the night tidepool tours conducted during the fall and winter.
FOR MORE INFORMATION: Write to: James V. Fitzgerald Marine Reserve, c/o County of San Mateo Parks & Recreation Commission, 400 Marshall Street, Redwood City, CA 94063. Tel: (415) 363-4020. Web site: http://www.prusik.com/pads/SanMateo/SanMateo.html

   Bean Hollow State Beach

SITE DESCRIPTION: Exposed coast with boulders and pocket sandy beaches.
NEAREST CITY: Pescadero.
ACCESS: Drive south on Highway 1. Pebble Beach is $17^{1/2}$ miles south of Half Moon Bay; 3 miles south of Pescadero.
NOTES: Hollow State Beach is noted for its fishing and picnicking, and for its distinctive tafoni rock formations.

The boulders at Pebble Beach, which have been shaped by erosion over the centuries, are always a source of beauty and wonder, as are the colorful pebbles in the area—truly nature's architecture in miniature.

This is also an excellent place to view seashore creatures. Several sea palms (see p. 232) are found on the rocks here, indicating the harsh nature of the coastline. Flat-bottomed periwinkles (p. 85), ribbed limpets (p. 77) and rough limpets (p. 78) may be seen side by side on the same rocks. If you see a black turban (p. 81), look closely to see tiny black limpets (p. 77) attached. Tidepools hold a variety of marine inhabitants here, including the opalescent nudibranch (p. 124) and Monterey dorid (p. 118).

FOR MORE INFORMATION: Write to: Bean Hollow State Beach, c/o San Francisco Bay Area District Headquarters, 250 Executive Park Boulevard, Suite 4900, San Francisco, CA 94134. Tel: (650) 879-2170. Web site: http://cal-parks.ca.gov/ *or* http://www.mbnms.nos.noaa.gov/Visitor/Access/bean.html

SANTA CRUZ COUNTY

   Natural Bridges State Beach

SITE DESCRIPTION: Exposed site with a sandy beach and rocky tidepool area.
NEAREST CITY: Santa Cruz.
ACCESS: From Highway 1, turn west at Swift Avenue; or from Santa Cruz, drive north on West Cliff Drive until it ends at the park.
NOTES: The wildlife at this beach includes spectacular migrations of the monarch butterfly. As many as 1,500,000 monarchs inhabit the area for part of each winter. Brown pelicans and shorebirds are commonly seen here, as are seals, otters and migrating whales. Marine life includes large numbers of sand-castle worms (see p. 50), as well as Hartweg's chiton (p. 61), California spiny chiton (p. 64) and red thatched barnacles (p. 163) attached to rocks. To reach the tidepools, walk north of the beach and proceed along the edge. Tidepools harbor giant green anemones (p. 30), purple stars (p. 197) and California mussels (p. 125). A visitor center located in the park has a variety of information, butterfly shirts, and a selection of books and postcards.

Be sure to visit the nearby Seymour Marine Discovery Center in Santa Cruz. It features many excellent live displays of central California marine life, and it stocks marine gifts, clothing, books and educational materials for all ages.

FOR MORE INFORMATION: Write to: Santa Cruz District and Sector Headquarters, 600 Ocean Street, Santa Cruz, CA 95060. Tel: (831) 423-4609 or (831) 459-3800 (Seymour Marine Discovery Center). Web site: http://cal-parks.ca.gov/

MONTEREY COUNTY

  Elkhorn Slough

SITE DESCRIPTION: Protected mud and sand shores.
NEAREST CITY: Moss Landing.
ACCESS: From Moss Landing, drive east on Dolan Road for about 4 miles, turn north on Elkhorn Road and travel 2 miles to reach the visitor center.
NOTES: Elkhorn Slough is a sheltered body of salt water almost 7 miles long. A portion of the area is open to the public for hiking and nature activities, and the Elkhorn Slough National Research Reserve and visitor center are located here. (Call ahead for hours of operation.)

Access to the seashore is via Kirby Park, 2 miles north of the visitor center on Elkhorn Road. Note that the tides here occur 25 minutes later than on the coast. This is an excellent area for viewing wildlife that lives in a protected area with a sand or mud shoreline. You may see the mudflat snail (see p. 86), Mediterranean mussel (p. 126), Atlantic oyster (p. 130), California jackknife clam (p. 141) and California lyonsia (p. 158). As well, brown pelicans, Caspian terns and great egrets inhabit this region.
FOR MORE INFORMATION: Write to: Elkhorn Slough National Estuarine Research Reserve, 1700 Elkhorn Road, Watsonville, CA 95076. Tel: (831) 728-2822. Web site: http://www.elkhornslough.org/

    Lover's Point

SITE DESCRIPTION: Semi-exposed site with a rocky area, boulders, tidepools and a sandy beach.
NEAREST CITY: Pacific Grove.
ACCESS: In Pacific Grove, on Ocean View Boulevard at Forest Avenue.
NOTES: This site is part of Pacific Grove Marine Refuge, where marine life is protected. Sea otters can often be seen resting among the floating kelp, and there are fabulous tidepools containing bat stars (see p. 193), white spotted sea goddess (p. 119) and hooked slippersnails (p. 89) clinging on the shells of black turbans (p. 81). The empty shells of kelp scallop (p. 132), appleseed erato (p. 92) and California trivia (p. 93) frequently wash up on the beach at Lover's Point.

This very popular site is used for a wide variety of activities such as kayaking, picnicking and strolling. Lover's Point is a little less than a mile west of the Monterey Bay Aquarium. This impressive facility features displays of marine life, including a wide variety of jellies, which are fascinating to watch underwater.
FOR MORE INFORMATION: Write to: City of Pacific Grove, Parks and Recreation Department, 300 Forest Avenue, Pacific Grove, CA 93950. Tel: (831) 648-3100. Web site: http://www2.ci.pacific-grove.ca.us/

Point Pinos

SITE DESCRIPTION: Exposed rocky shore with pocket beaches.
NEAREST CITY: Pacific Grove.
ACCESS: In Pacific Grove, on Ocean View Boulevard. Ample street parking is available.
NOTES: Point Pinos is a rugged, exposed area of the coast that is subject to strong waves and surf. It is a very productive intertidal site, part of Pacific Grove Marine Refuge, where all marine life is protected. Be sure to leave all creatures where you find them so that others may enjoy them. Many hearty marine invertebrates inhabit this site, including aggregating anemone (see p. 29), Hartweg's chiton (p. 61), mossy chiton (p. 68) and red thatched barnacle (p. 163).

The Point Pinos Lighthouse, the oldest continuously operated lighthouse on the west coast, is located on the opposite side of Ocean View Boulevard.
FOR MORE INFORMATION: Write to: City of Pacific Grove, Parks and Recreation Department, 300 Forest Avenue, Pacific Grove, CA 93950. Tel: (831) 648-3100 or (408) 372-2809. Web site: http://www2.ci.pacific-grove.ca.us/

Asilomar State Beach

SITE DESCRIPTION: Exposed, rocky site with tidepools and small sandy pocket beaches.
NEAREST CITY: Pacific Grove.
ACCESS: In Pacific Grove, travel south on Ocean View Boulevard and follow it until it changes into Sunset Drive along the coast.
NOTES: Located on the scenic Monterey Peninsula in Pacific Grove,

Asilomar State Beach is a prime tidepooling site with wild waves and hazardous rip currents. As in all exposed coastal sites, be careful here, and never turn your back to the water. Marine life at this site is protected as part of Pacific Grove Marine Refuge, but sport fishing is permitted. A few intertidal species to watch for here are branch-ribbed mussel (see p. 127), volcano keyhole limpet (p. 73), flat hoofsnail (p. 89) and the empty shells of black abalone (p. 71).

FOR MORE INFORMATION: Write to: Monterey District Headquarters, 2211 Garden Road, Monterey, CA 93940. Tel: (831) 642-4242. Web site: http://cal-parks.ca.gov/

Carmel River State Beach

SITE DESCRIPTION: Exposed sandy beach with a boulder edge.
NEAREST CITY: Carmel.
ACCESS: On Highway 1, immediately south of Carmel.
NOTES: Carmel River State Beach features a fabulous sandy beach and a rocky intertidal site at Monastery Beach. Watch for sea otters, or listen to hear them munching. Other marine life here includes the goose barnacle (see p. 165), giant Pacific chiton (p. 69) and Heath's chiton (p. 67).

The surf is extremely dangerous here, but the rough water washes an excellent variety of shells on the beach. These include black abalone (see p. 71), variegated amphissa (p. 103), brown turban (p. 82) and apple-seed erato (p. 92).

This park also features a bird sanctuary in the lagoon (just before the Carmel River empties into the sea), where you can see a variety of waterfowl and songbirds. You can reach the sanctuary directly from Carmel.

FOR MORE INFORMATION: Write to: Monterey District Headquarters, 2211 Garden Road, Monterey, CA 93940. Tel: (831) 649-2836. Web site: http://cal-parks.ca.gov/

Point Lobos State Reserve

SITE DESCRIPTION: Semi-exposed, rocky sites.
NEAREST CITY: Carmel.
ACCESS: From Carmel, drive 2 1/2 miles south on Highway 1.
NOTES: This popular rugged preserve includes a wide range of impressive coastal areas, with pocket beaches interspersed among rocky outcroppings. Marine mammals to be found here include sea otter, harbour seal, northern elephant seal and California sea lion. There are several trails that take hikers to various destinations, including Weston Beach, a great intertidal site along the south shore trail. The rugged coast within this reserve is dangerous—exercise caution at all times.

A few of the intertidal species to be seen here are the purple sea urchin (see p. 203), giant owl limpet (p. 79), California mussel (p. 125) and goose barnacle (p. 165), along with a large number of seaweeds. All waters surrounding this ecological reserve, and its animal and plant life, are protected. The reserve is also an excellent site for viewing sandstone and conglomerates and other interesting rocks.
FOR MORE INFORMATION: Write to: Point Lobos State Reserve, P.O. Box 62, Carmel, CA 93923. Tel: (831) 624-4909. Web site: http://www.pt-lobos.parks.state.ca.us/

Garrapata State Beach

SITE DESCRIPTION: Exposed site with a rocky shoreline and several large tidepools.
NEAREST CITY: Big Sur.
ACCESS: From Big Sur, drive 18 miles north on Highway 1. Take the Point Lobos State Reserve turnoff and travel 3 1/2 miles south. Several pullouts are available at spots along the park's coastline.
NOTES: Many marine life forms can be found at this fabulous site. Soberanes Point is an excellent area to view intertidal life, and at other locations to the north, large tidepools contain the aggregating anemone (see p. 29), giant green anemone (p. 30), painted anemone (p. 33), black abalone (p. 71), lined chiton (p. 58) and giant spined star (p. 196).

Wildlife observers can watch sea otters playing in the surf here, and gray whales are often spotted during their migration (November to May).

The cliffs in this park contain large middens with abalone shells, a traditional source of food for local aboriginal people.
FOR MORE INFORMATION: Write to: Monterey District Headquarters, 2211 Garden Road, Monterey, CA 93940. Tel: (831) 624-4909. Web site: http://cal-parks.ca.gov/

SAN LUIS OBISPO COUNTY

Cayucos State Beach

SITE DESCRIPTION: Semi-exposed, sandy beach with rock outcroppings.
NEAREST CITY: Morro Bay.
ACCESS: From Morro Bay, take Highway 1 north, drive 5 miles to Cayucos, and follow Cayucos Drive to the water.
NOTES: This picturesque site, a mixture of beautiful sand and large rocky outcroppings, attracts a wide variety of marine life. The intertidal area extends along the shore on both sides of a pier, which is lit for night fishing.

The creatures living in the sand include moonglow anemone (see p. 28), Pacific mole crab (p. 178) and purple olive (p. 107). In more rocky areas you may see the black turban (p. 81) and shell-binding colonial worms (p. 53).
FOR MORE INFORMATION: Write to: San Luis Obispo Coast District Headquarters & Sector Headquarters, 3220 South Higuera Street, Suite 311, San Luis Obispo, CA 93401. Tel: (805) 781-5200. Web site: http://cal-parks.ca.gov/

Montaña de Oro State Park

SITE DESCRIPTION: Exposed site with rock benches and tidepools of various sizes.
NEAREST CITY: San Luis Obispo.
ACCESS: From San Luis Obispo, drive north on Highway 101, take the Los Osos exit and drive west on Los Osos Valley Road for 12 miles to the park entrance.
NOTES: This state park is a treasure, with more than 8,000 acres of remote beaches, beautiful coves, rugged cliffs, canyons and much more. It is ideal for many pursuits, including hiking, horseback riding, camping and picnicking.

Coral Cove is an intertidal site that should not be missed. A short hike along the Bluff Trail takes you there, and you can observe a great variety of marine creatures in the many tidepools. Strawberry anemone (see p. 35), giant green anemone (p. 30), yellow-gilled sea goddess (p. 119), umbrella crab (p. 174) and giant spined star (p. 196) are just a few local species, and many marine algae can also be seen here. Harbor seals can often be seen sitting high on the rocks, and sea otters are regular visitors in the waters below.

Be sure to stop in at the visitor center located at park headquarters.

FOR MORE INFORMATION: Write to: San Luis Obispo Coast District Headquarters & Sector Headquarters, 3220 South Higuera Street, Suite 311, San Luis Obispo, CA 93401. Tel: (805) 528-0513 or 772-7434. Web site: http://cal-parks.ca.gov/

Pismo State Beach

SITE DESCRIPTION: Exposed sandy beach.
NEAREST CITY: Oceano.
ACCESS: From Highway 101 in Pismo Beach, drive south on Dolliver Street (Highway 1) and turn west at Pier Avenue.
NOTES: Pismo State Beach and the adjacent recreation area is a fabulous sandy beach where you can spend many hours walking along the seashore. Wave after wave roll onto the southern section of beach, often exposing Pismo clams (see p. 151) and bean clams (p. 144), and sometimes washing the purple-striped jelly (p. 25) onto the shore.

Look in the sand at low tide and you may find the purple olive (see p. 107), zigzag olive (p. 108), Pacific mole crab (p. 178), eccentric sand dollar (p. 204) and moonsnail hermit (p. 172) living in the empty shells of Lewis's moonsnail (p. 94). The largest overwintering concentrations of monarch butterflies in California can be found at North Beach campground during the winter months. This is truly a remarkable sight!

Camping, swimming, surf fishing and picnicking beside the water are popular activities in this park. Be sure to stop in at the nature center at Oceano Campground.
FOR MORE INFORMATION: Write to: San Luis Obispo Coast District Headquarters & Sector Headquarters, 3220 South Higuera Street, Suite 311, San Luis Obispo, CA 93401. Tel: (805) 549-3312 or (805) 489-2684. Web site: http://cal-parks.ca.gov/

SANTA BARBARA COUNTY

   Refugio State Beach

SITE DESCRIPTION: Exposed site with a mix of rock and sandy beaches.
NEAREST CITY: Santa Barbara.
ACCESS: From Santa Barbara, take Highway 101 and drive northwest for 23 miles.
NOTES: At this beach you can camp under palm trees, and go kayaking, boogie boarding and/or suntanning. A bike trail along the bluff connects this beach with El Capitán State Beach 2 1/2 miles to the east.

Located south of Point Conception, this park is the perfect habitat for a wide range of marine life. California spiny lobster (see p. 171), black turban (p. 81), wavy turban (p. 81), chestnut cowrie (p. 93), western sea roach (p. 167) and scale-sided piddock (p. 155) are a few of the species to be found here. The presence of California mussel (p. 125) and goose barnacle (p. 165) indicate the rough surf conditions of the area. Remember to treat the ocean with respect and always keep a close eye on the water.
FOR MORE INFORMATION: Write to: Channel Coast District Headquarters, 1933 Cliff Drive, Suite 27, Santa Barbara, CA 93109. Tel: (805) 968-1033. Web site: http://cal-parks.ca.gov/

    El Capitán State Beach

SITE DESCRIPTION: Exposed sandy beach with boulders interspersed with tidepools.
NEAREST CITY: Goleta.
ACCESS: From Goleta, travel northwest on Highway 101 for 8 miles and follow the signs.
NOTES: This beach is an excellent place to go for a swim or try surfing, surf fishing, picnicking or camping. Here sicamore oak stand alongside El Capitán Creek. The rocks in this area are occasionally covered with tar. This is a local phenomenon—tar oozes from the ground naturally. The Channel Islands can be viewed from El Capitán Point.

The tidepools are adjacent to the campground at the day use area. Watch for black turban (see p. 81), black limpet (p. 77), giant owl limpet (p. 79), volcano keyhole limpet (p. 73), spot-bellied rock crab (p. 183) and purple star (p. 197).
FOR MORE INFORMATION: Write to: Channel Coast District Headquarters, 1933 Cliff Drive, Suite 27, Santa Barbara, CA 93109. Tel: (805) 899-1400. Web site: http://cal-parks.ca.gov/

Carpinteria State Beach

SITE DESCRIPTION: Semi-exposed site with a sandy beach, a rocky reef and tidepools.
NEAREST CITY: Carpinteria.
ACCESS: From Santa Barbara, take the downtown Carpinteria exit or state beach exit and drive 12 miles south, following the signs.
NOTES: A visit to Carpinteria State Beach is a great opportunity to camp beside a fabulous beach and tidepool area. This is a very popular park with campers, so make reservations. A visitor center is open daily with educational displays and a live intertidal exhibit.

Low tide brings a variety of marine life into view, including the wavy turban (see p. 81), Kellet's whelk (p. 104), California sea hare (p. 112), stiff-footed sea cucumber (p. 206) and long-stalked sea squirt (p. 209). A variety of shells also wash up on the sandy beach, among them California hornsnail (p. 87), hooked slippersnail (p. 89), gray snakeskin-snail (p. 110), secret jewelbox (p. 135), Pacific calico scallop (p. 132) and boring softshell-clam (p. 152).

Harbor seals live in the intertidal zone as well. Please keep your distance when watching them bask on the rocks, and don't approach or disturb them.
FOR MORE INFORMATION: Write to: Channel Coast District Headquarters, 1933 Cliff Drive, Suite 27, Santa Barbara, CA 93109. Tel: (805) 684-2811 or (805) 968-3294. Web site: http://cal-parks.ca.gov/

VENTURA COUNTY

  Channel Islands National Park

Santa Cruz Island.

SITE DESCRIPTION: Exposed rocky sites.
NEAREST CITY: Ventura.
ACCESS: Visitors may take a boat from Ventura or Oxnard, or take a plane from Camarillo or Santa Barbara. Contact the park for transportation information.
NOTES: This fabulous island preserve is being restored to its original state. Mountains, sandy beaches, interesting coves and sea caves are yours to explore in this wild area. Tidepools form in many locations, but the most accessible ones are on Anacapa Island (Frenchys Cove) and Santa Barbara Island. Intertidal invertebrates present include orange cup coral (see p. 28), purple star (p. 197) and purple sea urchin (p. 203). All marine life here is protected—you may not take specimens away.

Recreation activities include kayaking, hiking and boating. Camping is permitted but permits and reservations are required. An excellent visitor center located in Ventura has live invertebrates and touch tanks.

FOR MORE INFORMATION: Write to: Superintendent, Channel Islands National Park, 1901 Spinnaker Drive, Ventura, California 93001. Tel: (805) 658-5730. Web site: http://www.nps.gov/chis/

SOUTH COAST

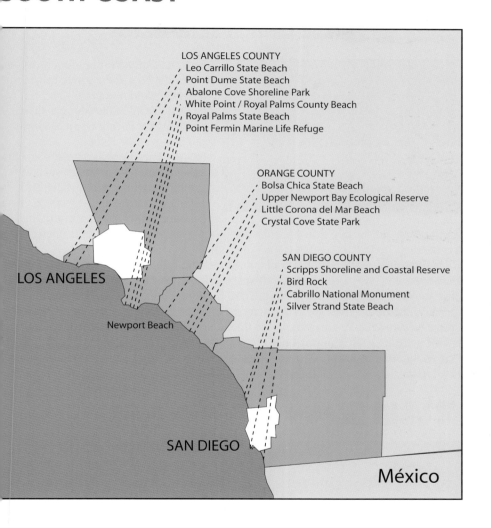

LOS ANGELES COUNTY

    Leo Carrillo State Beach

SITE DESCRIPTION: Exposed, rocky site with tidepools and a sandy beach.
NEAREST CITY: Malibu.
ACCESS: From Santa Monica, drive north on the Pacific Coast Highway (Highway 1) for 28 miles.
NOTES: This beach is a great place to enjoy a variety of activities such as camping, surfing, windsurfing, surf fishing, jogging and just relaxing. Two beautiful sandy beaches, north and south beach, are located near the entrance to the park. A visitor center is open on weekends.

At the south beach area, tidepool enthusiasts can find species such as the moonglow anemone (see p. 28), banded turban (p. 83), angular unicorn (p. 95), purple olive (p. 107), Poulson's rock snail (p. 101), California cone (p. 109), kelp scallop (p. 132) and hairy hermit (p. 173). Intertidal life is best viewed at low minus tides. Please be sure to leave creatures in the tidepools. Wildlife to watch here includes brown pelicans and California quail.

FOR MORE INFORMATION: Write to: Angeles District Headquarters, 1925 Las Virgenes Road, Calabasas, CA 91302. Tel: (818) 880-0350. Web site: http://cal-parks.ca.gov/

    Point Dume State Beach

SITE DESCRIPTION: Exposed rocky shores with sandy beaches and tidepools.
NEAREST CITY: Malibu.
ACCESS: From Santa Monica, take Highway 1 west for 18 miles to Westward Road.
NOTES: Fabulous views are a highlight at Point Dume State Beach, as well as excellent viewing points for gray whales migrating through this area between November and May. Opportunities abound for swimming, surfing, scuba diving, rock climbing, surf fishing and lying on the beach.

Landscape features here include headlands, cliffs, secluded coves with large boulders, and tidepools that border the beaches. Marine life to be observed includes the aggregating anemone (see p. 29), sand-castle worm (p. 50), California mussel (p. 125), goose barnacle (p. 165) and striped shore crab (p. 185).

FOR MORE INFORMATION: Write to: Angeles District Headquarters, 1925 Las Virgenes Road, Calabasas, CA 91302. Tel: (310) 457-8143. Web site: http://cal-parks.ca.gov/

 Abalone Cove Shoreline Park

SITE DESCRIPTION: Semi-exposed rocky site with tidepools and a sandy beach.
NEAREST CITY: Palos Verdes.
ACCESS: The park is in Palos Verdes, on Polo Verdes Drive South (an extension of 25th Street).
NOTES: This rich intertidal site includes an ecological reserve, thus no marine life may be removed from the area. The park is operated by the City of Palos Verdes, which charges a fee to park at the site. From the parking lot, visitors walk down a short trail to the seashore and tidepools. Marine life here includes giant keyhole limpet (see p. 74), Kellet's whelk (p. 104), California sea hare (p. 112), red thatched barnacle (p. 163), bat star (p. 193), purple sea urchin (p. 203) and red sea urchin (p. 203).

Other favorite activities here include picnicking, fishing, diving and just relaxing on the secluded sandy beach. Listen for the sound of the "laughing pebbles" when receding waves ripple over the rocks.

FOR MORE INFORMATION: Write to: Abalone Cove Shoreline Park and Ecological Reserve, 5970 Palos Verdes Drive South, Rancho Palos Verdes, CA 90275. Tel: (310) 377-1222.

White Point/Royal Palms County Beach

SITE DESCRIPTION: Exposed rocky site with tidepools.
NEAREST CITY: Palos Verdes.
ACCESS: The beach is in Palos Verdes, on Polo Verdes Drive just south of the Western Avenue turnoff.
NOTES: At Royal Palms County Beach, local fishermen wet their lines and beachgoers visit a small sandy beach that is exposed at low tide. The remains of an old bathhouse can be seen on the rocks. At one time the hot springs were heated by offshore geothermal vents.

There is easy access to many small rockbound tidepools here, which are clear enough for good viewing and photography. Marine life to see at this site includes giant green anemone (see p. 30), bat star (p. 193), purple star (p. 197), purple sea urchin (p. 203), wavy turban (p. 81), black turban (p. 81) and giant keyhole limpet (p. 74).
FOR MORE INFORMATION: Write to: City of Los Angeles Recreation and Parks Department, White Point Park, Paseo Del Mar and Western Avenue, San Pedro, CA 90731. Web site: http://www.cityofla.org

Royal Palms State Beach

SITE DESCRIPTION: Semi-exposed rocky site with tidepools.
NEAREST CITY: Palos Verdes.
ACCESS: The beach is in Palos Verdes, on Polo Verdes Drive just south of the Western Avenue turnoff.
NOTES: Majestic palms and a fabulous coastline make Royal Palms State Beach a spectacular park, especially when viewed from the top of the bench. It is a quiet state park that adjoins Royal Palms County Beach (above). Tidepoolers and picnickers frequent the day use area. Street parking is available.

A paved walkway and stairway lead to the intertidal area below the bench. This site is somewhat silty and includes mossy chiton (see p. 68), chestnut cowrie (p. 93), California cone (p. 109), bay ghost shrimp (p. 171), blueband hermit (p. 173), bat star (p. 193), purple sea urchin (p. 203), two-spotted octopus (p. 159) and western spiny brittle star (p. 200).

FOR MORE INFORMATION: Write to: Royal Palms State Beach, 2000 W. Paseo Del Mar, San Pedro, CA 90732-4724. Web site: http://www.sanpedrochamber.com/champint/roylplms.htm

Point Fermin Marine Life Refuge

SITE DESCRIPTION: Exposed rocky site with a sand and mud mixture.
NEAREST CITY: San Pedro.
ACCESS: From the San Diego Freeway, turn south on the Harbor Freeway (Highway 110) and follow it to 22nd Street (past the point at which it becomes Gaffey Street). Turn east on 22nd Street and follow the signs to Cabrillo Marine Aquarium. Parking is at the aquarium or on the street.

NOTES: This area is rich in marine life. Here large rocks, boulders and tidepools hold a variety of interesting creatures, including California cone (see p. 109), chestnut cowrie (p. 93), white spotted sea goddess (p. 119), green false-jingle (p. 134), scale-sided piddock (p. 155), flat-tip piddock (p. 155), two-spotted octopus (p. 159) and flat-spined brittle star (p. 201).

Be sure to stop in at the Cabrillo Marine Aquarium, which has many interesting displays and a wide variety of local marine life in aquariums. As well, it offers periodic guided tidepool walks at Point Fermin and scheduled night walks from March through July to observe the amazing phenomenon of California grunion spawning.

FOR MORE INFORMATION: Write to: Cabrillo Marine Aquarium, 3720 Stephen White Drive, San Pedro, CA 90731. Tel: (310) 548-7562. Web site: http://www.cabrilloaq.org/

ORANGE COUNTY

  Bolsa Chica State Beach

Pacific calico scallop.

SITE DESCRIPTION: Exposed sandy beach.
NEAREST CITY: Huntington Beach.
ACCESS: The beach is located in Huntington Beach, on Highway 1 between Warner Avenue and Golden West Street.
NOTES: Bolsa Chica State Beach is a popular destination for people interested in surf fishing, surfing, camping, picnicking—and beachcombing, with 2 miles of beach to explore. Sunset watching is also a popular pastime among both locals and visitors.

The beach is not an intertidal site but rather an excellent location to walk the beach and search for the many shells that are washed up on shore, including spiny cup-and-saucer (see p. 91), secret jewelbox (p. 135), Pacific calico scallop (p. 132), pearly jingle (p. 133), Pacific eggcockle (p. 137), wavy venus (p. 147) and white venus (p. 146). You will get the best beachcombing results on a receding tide.

The Bolsa Chica Ecological Reserve is located on the east side of Highway 1.
FOR MORE INFORMATION: Write to: Bolsa Chica State Beach, 17851 Pacific Coast Highway, Huntington Beach, CA 92642. Tel: (714) 846-3460. Web site: http://cal-parks.ca.gov/

  Upper Newport Bay Ecological Reserve

SITE DESCRIPTION: Protected mud estuary.
NEAREST CITY: Newport Beach.
ACCESS: From Newport Beach, drive north on Jamboree Road. Turn west at Black Bay Drive and follow the road for a short distance to reach the reserve.
NOTES: This ecological reserve offers an excellent opportunity to view many of the creatures that live in an estuary environment. Fewer and fewer such areas exist today, due to development. The area is also ideal for a wide range of recreational pursuits such as hiking, biking, fishing and canoeing. It is also a well-known destination for birders.

Marine invertebrates here include the mudflat snail (see p. 86), California bubble (p. 110), Mediterranean mussel (p. 126), striped shore crab (p. 185) and Mexican fiddler (p. 187).

FOR MORE INFORMATION: Write to: Upper Newport Bay Naturalists, 600 Shellmaker Drive, Newport Beach, CA 92660. Tel: (949) 640-6746. Web site: http://newportbay.org/

    ## Little Corona del Mar Beach

SITE DESCRIPTION: Exposed sandy beach with a rocky outcropping and tidepools.
NEAREST CITY: Corona del Mar.
ACCESS: The trail to the beach starts at Ocean Boulevard and Poppy Avenue in Corona del Mar.
NOTES: Sunbathing and tidepooling are popular activities at this fabulous beach, easily accessible via a well-traveled walkway from the street.
Intertidal life is protected here and in the surrounding area, so no marine life can be moved or taken out of the site without a scientific permit. The rocky tidepools here harbor the sand-castle worm (see p. 50), blueband hermit (p. 173), striped shore crab (p. 185), bat star (p. 193), purple star (p. 197), purple sea urchin (p. 203), red sea urchin (p. 203) and banded brittle star (p. 200), among other species. Like most tidepools, these windows into the sea are fascinating if you wait quietly for a while to see what interesting species come out of hiding when the coast seems clear.

FOR MORE INFORMATION: Write to: City of Newport Beach, P.O. Box 1768, Newport Beach, CA 92658-8915. Tel: Newport Beach Marine Department (714) 644-3044. Web site: http://www.newportbeachonline.com/

Crystal Cove State Park

SITE DESCRIPTION: Semi-exposed rocky coast with sandy beaches.
NEAREST CITY: Laguna Beach.
ACCESS: The park is on the Pacific Coast Highway, immediately South of Corona Del Mar.
NOTES: Crystal Cove State Park is a large wilderness park with 2,000 acres of undeveloped woodlands and 3 1/2 miles of beach, where visitors can engage in hiking, horseback riding, surfing, diving, fishing and swimming.

The Pelican Point site, at the north end of the park, is an excellent area for tidepooling. It is accessible from the parking area via a short walk. Beachcombers here can discover the giant keyhole limpet (see p. 74), California sea hare (p. 112), California black sea hare (p. 113), green bubble (p. 111) and two-spotted octopus (p. 159). Wildlife in the park includes mule deer, badgers and red-tailed hawks. For additional information about the park, stop by the ranger station.
FOR MORE INFORMATION: Write to: Crystal Cove State Park, 8471 Pacific Coast Highway, Laguna Beach, CA 92651. Tel: (949) 492-3539.
Web site: http://cal-parks.ca.gov/

SAN DIEGO COUNTY

San Elijo State Beach

SITE DESCRIPTION: Semi-exposed, sandy beach with rocks and tidepools.
NEAREST CITY: Cardiff-by-the-Sea.
ACCESS: From the entrance channel to San Elijo Lagoon, near the community of Cardiff-by-the-Sea, drive north on Highway 101 for 3/4 mile.
NOTES: At this beach you can camp near the intertidal area, or park at the day use area on the street and walk south to the tidepools below the campground. The park also offers opportunities for swimming, surfing, snorkeling and picnicking.

The tidepools here hold several interesting finds, including the wavy turban (see p. 81), California sea hare (p. 112) and two-spotted octopus (p. 159). Some of the empty

shells you may find on the beach include festive murex (p. 98), kelp scallop (p. 132), secret jewelbox (p. 135), boring softshell-clam (p. 152) and rough piddock (p. 156).
FOR MORE INFORMATION: Write to: California Department of Parks & Recreation, San Diego Coast District Headquarters, 9609 Waples Street, Suite 200, San Diego, CA 92121. Tel: (619) 642-4200 or (760) 753-5091. Web site: http://cal-parks.ca.gov/

   ## Scripps Shoreline and Coastal Reserve

Scripps Shoreline from Scripps Pier.

SITE DESCRIPTION: Exposed sandy beach with rocks and boulders.
NEAREST CITY: La Jolla.
ACCESS: From the San Diego Freeway 5, take Ardath Road and drive west for about 3 miles, then turn north onto La Jolla Shores Drive for 2 miles. Park on the street.
NOTES: Scripps Shoreline is also known as Dike Rock, which takes its name from the geological term for lava flowing through layered beach rocks. Scripps beach is on the north side of the landmark pier at La Jolla. All marine life is protected in the University of California Scripps Coastal Reserve, including the intertidal area and waters extending 1,000 feet.

The variety of intertidal life here includes angular unicorn (see p. 95), checkered unicorn (p. 96), Poulson's rock snail (p. 101), festive murex (p. 98), California cone (p. 109), California sunset clam (p. 145) and banded brittle star (p. 200).

This area is popular for diving, sunbathing, surfing, surf fishing and other activities. At the Birch Aquarium at Scripps Institution of Oceanography, located nearby, visitors can view a wide variety of marine life from around the world, see educational exhibits and browse in a bookstore.

FOR MORE INFORMATION: Write to: Scripps Institution of Oceanography, University of California San Diego, La Jolla, CA 92093-0233. Tel: (858) 534-6324. Web site: http://www.sio.ucsd.edu

Bird Rock

SITE DESCRIPTION: Exposed rocky site with tidepools.
NEAREST CITY: La Jolla.
ACCESS: From La Jolla Boulevard, turn west at Bird Rock Avenue and drive to the end. Parking is on the street.
NOTES: Bird Rock has no official status as a park. It is the local name for a large rock connected to shore at low tide and inhabited by a variety of birds (and, as a result, extensively whitewashed). The area is also favored by surfers. A short set of stairs and a quick scramble over rubble takes the visitor to the intertidal area. Keep your eye on the water at this wave-prone site.

A few of the interesting creatures found attached to the sandstone or in tidepools here include the giant green anemone (see p. 30), aggregating anemone (p. 29), giant owl limpet (p. 79), banded turban (p. 83), gilded turban (p. 84) and California spiny lobster (p. 171). A minus tide will uncover the widest assortment of marine life.

  ## Cabrillo National Monument

SITE DESCRIPTION: Semi-exposed, rocky area with many large tidepools.
NEAREST CITY: San Diego.
ACCESS: From San Diego, take Catalina Boulevard and Cabrillo Memorial Drive south to the entrance of the national monument.
NOTES: Cabrillo National Monument is an important natural preserve for a wide range of wildlife in southern California. Several miles of hiking trails take visitors through this impressive preserve. The visitor center is an excellent place to learn about the park and its inhabitants through displays, books and visual presentations, and from here there are superb views of the harbor and surrounding area. The Point Loma Lighthouse is located nearby.

The tidepools at this outstanding site, one of the few remaining tidepool areas in southern California, are fabulous. You can get to them via a side road just before the formal entrance to the monument. All plants, animals and shells here are protected, so nothing can be removed. Visitors may discover green abalone (see p. 30), giant

keyhole limpet (p. 74), chestnut cowrie (p. 93), California cone (p. 109), California black sea hare (p. 113), globose kelp crab (p. 180) and bat star (p. 193). This is a place where it is easy to forget about the passing of time.
FOR MORE INFORMATION: Write to: Cabrillo National Monument, 1800 Cabrillo Memorial Drive, San Diego, CA 921106. Tel: (619) 557-5450.
Web site: http://www.nps.gov/cabr/vc.html

 ## Silver Strand State Beach

SITE DESCRIPTION: Exposed sandy beach.
NEAREST CITY: Coronado.
ACCESS: From Coronado, travel 4 1/2 miles south on Highway 75. Turn west at Silver Strand Boulevard, where the park is located.
NOTES: Silver Strand State Beach offers the perfect opportunity to take your shoes off and go for a stroll on a deluxe beach. Bring the suntan lotion—this sand spit offers 2 1/2 miles of ocean beach and half a mile of bay frontage. Surfing, surf fishing and boogie boarding are popular activities at this park.

Silver Strand is known for the large numbers of shells deposited on its beautiful beach. Some of the mollusks or their shells that can often be found here are spiny cup-and-saucer (see p. 91), Pacific calico scallop (p. 132), pearly jingle (p. 133), Pismo clam (p. 151) and bean clam (p. 144), often with clam hydroids (p. 21) attached. Other species that may be found here include red-striped acorn barnacle (p. 164) attached to an eccentric sand dollar (p. 204), or white-ribbed red barnacle (p. 164) on the shell of a California mussel (p. 125).

California grunion (p. 217) also make their spectacular night run at this site.
FOR MORE INFORMATION: Write to: Silver Strand State Beach, 5000 Highway 75, Coronado, CA 92118. Tel: (619) 435-5184. Web site: http://cal-parks.ca.gov/

Acknowledgements

I would like to thank the many people who assisted with this project in so many ways and without whose help the book could not have been completed.

Thanks to Howard White of Harbour Publishing for accepting this book, as well as Peter Robson for all his help in coordinating the project, Martin Nichols for his excellent layout and design, and Mary Schendlinger for her patient editing.

Special thanks to Andy Lamb, Vancouver Public Aquarium, Vancouver, BC, for his help in so many ways: identifying slides and confirming species, locating specimens, providing scientific editing, and much more.

Thanks also to these specialists for identifying and confirming species, locating specimens, providing scientific editing, forwarding reprints or species accounts, and making suggestions for the text: Bill Austin (Marine Ecology Station, Sidney, BC), Roger N. Clark (Klamath Falls, OR), Matthew Dick (Middlebury College, Middlebury, VT), Doug Eernisse (California State University, Fullerton, CA), Lindsey T. Groves (Natural History Museum of Los Angeles County, Los Angeles, CA), Rick Harbo (Fisheries & Oceans, Nanaimo, BC), Leslie H. Harris (Los Angeles County Museum of Natural History, Los Angeles, CA), Mike Hawkes (University of British Columbia, Vancouver, BC), John Holleman (San Andreas, CA), Gregory C. Jensen (University of Washington, Seattle, WA), Gretchen Lambert (Seattle, WA), Philip Lambert (Royal BC Museum, Victoria, BC), John C. Ljubenkov (Pauma Valley, CA), Val Macdonald (Biologica, Victoria, BC), Jim McLean (Los Angeles County Museum of Natural History, Los Angeles, CA), Sandra V. Millen (University of British Columbia, Vancouver, BC), Julie Oliveira (University of British Columbia, Vancouver, BC), Kenton Parker (Elkhorn Slough National Estuarine Research Reserve, CA), Pamela Roe (California State University, Stanislaus, Turlock, CA).

My appreciation to several institutions that kindly assisted with this project: Cabrillo Marine Aquarium (San Pedro, CA) allowed some specimens to be photographed in their aquariums. Jellies were photographed in the fabulous jellies section at the Monterey Aquarium, which also provided site information. Peter Macht (Seymour Marine Discovery Center, Santa Cruz, CA) generously provided several live specimens for photography. A few specimens from the Royal BC Museum, Victoria, BC, were photographed with the kind help of Philip Lambert.

Further Reading

Abbott, I.A., and G.J. Hollenberg. 1976. *Marine Algae of California*. Stanford CA: Stanford Univ. Press.

Behrens, David W. 1991. *Pacific Coast Nudibranchs: A Guide to the Opisthobranchs, Alaska to Baja*. Monterey CA: Sea Challengers.

California Coastal Commission. 1997. *California Access Guide* (rev. ed.). Berkeley: University of California Press.

Coan, Eugene V. et al. 2000. *Bivalve Seashells of Western North America: Marine Bivalve Mollusks from Arctic Alaska to Baja California*. Monograph #2. Santa Barbara CA: Santa Barbara Museum of Natural History.

Dawson, E.Y. 1982. *Seashore Plants of California*. California Natural History Guide #47. Berkeley: University of California Press.

Fitch, John E., and R.J. Lavenberg. 1975. *Tidepool and Nearshore Fishes of California*. Natural History Guide #38. Berkeley: University of California Press.

Harbo, R.M. 1997. *Shells & Shellfish of the Pacific Northwest: A Field Guide*. Madeira Park BC: Harbour Publishing.

Harbo, R.M. 1999. *Whelks to Whales: Coastal Marine Life of the Pacific Northwest*. Madeira Park BC: Harbour Publishing.

Hartman, O. 1968. *Atlas of the Errantiate Polychaetous Annelids from California*. Los Angeles: Alan Hancock Foundation, University of Southern California.

Hinton, Sam. 1987. *Seashore Life in Southern California: An Introduction to the Animal Life of California Beaches South of Santa Barbara, California* (rev. ed.). California Natural History Guide #26. Los Angeles: University of California Press.

Howorth, Peter C. 1978. *The Abalone Book*. Happy Camp CA: Naturegraph Publishers.

Jensen, Gregory C. 1995. *Pacific Coast Crabs & Shrimps*. Monterey CA: Sea Challengers.

Lambert, Philip. 2000. *Sea Stars of British Columbia, Southeast Alaska, and Puget Sound*. Vancouver: UBC Press.

McLean, James H. 1978. *Marine Shells of Southern California* (rev. ed.). Science Series 24, Zoology #11. Los Angeles: Natural History Museum of Los Angeles County.

Morris, R.H. et al. 1980. *Intertidal Invertebrates of California*. Stanford CA: Stanford University Press.

Niesen, Thomas M. 1994. *Beachcomber's Guide to California Marine Life*. Houston: Gulf Publishing Co.

Reish, Donald J. 1995. *Marine Life of Southern California: Emphasizing Marine Life of Los Angeles–Orange Counties* (2nd ed.). Dubuque IA: Kendall/Hunt Publishing Co.

Ricketts, Edward F., and Jack Calvin. 1985. *Between Pacific Tides* (rev. ed.) Stanford, CA: Stanford University Press.

Smith, R.I., and James T. Carlton, eds. 1975. *Light's Manual: Intertidal Invertebrates of the Central California Coast* (3rd ed.). Berkeley CA: University of California Press.

Tway, Linda E. 1991. *Tidepools Southern California: An Illustrated Guide to 100 Locations—from Pt. Conception to Mexico*. Santa Barbara CA: Capra Press.

Index

abalone, 71–72
Abalone Cove Shoreline Park, 279
abalone jingle shell (green false-jingle *Pododesmus macroschisma*), 134
Abarenicola claparedi (rough-skinned lugworm). *See under* Pacific lugworm (*Abarenicola pacifica*)
Abarenicola pacifica (Pacific lugworm), 49
aboriginal people. *See* Native people
Acanthina paucilirata (checkered unicorn), 96
Acanthina punctulata (spotted unicorn), 96
Acanthina spirata (angular unicorn), 95. *See also* spotted unicorn (*Acanthina punctulata*)
Acanthodoris columbina (rufus tipped nudibranch *Acanthodoris nanaimoensis*), 114
Acanthodoris nanaimoensis (rufus tipped nudibranch), 114
acid seaweed (flat acid kelp *Desmarestia ligulata*), 228
Acmaea asmi. *See* black limpet (*Lottia asmi*)
Acmaea insessa. *See* seaweed limpet (*Discurria insessa*)
Acmaea instabilis (unstable limpet *Lottia instabilis*), 79
Acmaea mitra (whitecap limpet), 75
Acmaea scabra. *See* rough limpet (*Lottia scabra*)
acorn barnacle (*Balanus glandula*), 163
Acrosiphonia coalita (tangle weed), 224
Adula californiensis (California datemussel), 128
Aeolidia papillosa (shag-rug nudibranch), 123
Aequipecten aequisulcatus (Pacific calico scallop *Argopecten ventricosus*), 132
Aequipecten circularis (Pacific calico scallop *Argopecten ventricosus*), 132
Aequorea aequorea (water jelly *Aequorea* sp.), 23
Aequorea sp. (water jelly), 23
Aequorea victoria (water jelly *Aequorea* sp.), 23
Agardhiella tenera. *See* succulent seaweed (*Sarcodiotheca gaudichaudii*)
Agassiz's peanut worm (*Phascolosoma agassizii*), 55
agate chama (secret jewelbox *Chama arcana*), 135
aggregating anemone (*Anthopleura elegantissima*), 29
Aglaophenia sp. (ostrich-plume hydroid), 23
Agriodesma saxicola (rock entodesma *Entodesma navicula*), 157
alabaster nudibranch (frosted nudibranch *Dirona albolineata*), 121
Alaria marginata (winged kelp), 231
alaria (winged kelp *Alaria marginata*), 231
Aletes squamigerus. *See* scaly tube snail (*Serpulorbis squamigerus*)
Alia carinata (carinate dovesnail), 102
Alloioplana californica. *See* oval flatworm (*Pseudoalloioplana californica*)
Amaroucium aequalisiphonis (multi-lobed tunicate *Ritterella aequalisiphonis*), 213
Amaroucium californicum. *See* California sea pork (*Aplidium californicum*)
Amaroucium solidum. *See* red sea pork (*Aplidium solidum*)
American glasswort (common pickleweed *Salicornia virginica*), 251, 252
American oyster (Atlantic oyster *Crassostrea virginica*), 130
Amiantis callosa (white venus), 146
Amphiodia occidentalis (long-armed brittle star), 199
Amphipholis squamata (dwarf brittle star), 199
Amphiporus bimaculatus (two-spotted ribbon worm), 41
Amphiporus formidabilis (white intertidal ribbon worm). *See under* pink-fronted ribbon worm (*Amphiporus imparispinosus*)
Amphiporus imparispinosus (pink-fronted ribbon worm), 42
Amphissa columbiana (wrinkled amphissa), 103
Amphissa versicolor (variegated amphissa), 103
amphipods, 168
ancient hoof shell (flat hoofsnail *Hipponix cranioides*), 89
angular unicorn (*Acanthina spirata*), 95
Anisodoris nobilis (sea lemon), 118
Annelida (phylum), 43–54
Anomia peruviana (pearly jingle), 133
Anoplarchus purpurescens (high cockscomb), 214, 218
Anthophyta (phylum), 251–53
Anthopleura artemisia (moonglow anemone), 28
Anthopleura elegantissima (aggregating anemone), 29
Anthopleura xanthogrammica (giant green anemone), 30
Anthozoa (class), 27–35
Antisabia cranioides (flat hoofsnail *Hipponix cranioides*), 89
Aphrodita japonica (copper-haired sea mouse), 46
Aphrodita refulgida. *See under* copper-haired sea mouse (*Aphrodita japonica*)
Aphrodite japonica. *See* copper-haired sea mouse (*Aphrodita japonica*)
Aplidium californicum (California sea pork), 210
Aplidium solidum (red sea pork), 211
Aplysia californica (California sea hare), 112
Aplysia vaccaria (California black sea hare), 113
apple seed erato (appleseed erato *Hespererato vitellina*), 92
apple-seed erato (appleseed erato *Hespererato vitellina*), 92
appleseed erato (*Hespererato vitellina*), 92
Aptyxis luteopicta. *See* painted spindle (*Fusinus luteopictus*)
Arachnida (class), 187
arachnids, 187
Archidistoma ritteri. *See* yellow lobed tunicate (*Eudistoma ritteri*)
Archidoris montereyensis (Monterey dorid), 118
Arctic hiatella (*Hiatella arctica*), 154
Arctic rock borer (Arctic hiatella *Hiatella arctica*), 154
Arctonoe fragilis (fragile scale-worm), 47

Arctonoe pulchra (red commensal scaleworm), 47
Arctonoe vittata (red-banded commensal scaleworm), 48
Argopecten ventricosus (Pacific calico scallop), 132
Arthropoda (phylum), 161–87
arthropods, 161–87
Asilomar State Beach, 269–70
Asterina miniata (bat star), 193
Asteroidea (class), 193–202
Astraea undosa. See wavy turban (*Megastraea undosa*)
Atlantic oyster (*Crassostrea virginica*), 130
Atlantic oyster drill (*Urosalpinx cinerea*), 102
Atlantic ribbed mussel (ribbed mussel *Geukensia demissa*), 127
Aurelia aurita. See under moon jelly (*Aurelia labiata*)
Aurelia labiata (moon jelly), 26
Axiognathus squamatus (dwarf brittle star *Amphipholis squamata*), 199
Axiothella rubrocincta (red-banded bamboo worm), 49
baetic olive (*Olivella baetica*), 107

Balanophyllia elegans (orange cup coral), 28
Balanus amphitrite (little striped barnacle). See under white-ribbed red barnacle (*Megabalanus californicus*)
Balanus glandula (acorn barnacle), 163
Balanus glandulus (acorn barnacle *Balanus glandula*), 163
Balanus pacificus (red-striped acorn barnacle), 164
Balanus tintinnabulum. See white-ribbed red barnacle (*Megabalanus californicus*)
Baltic macoma (*Macoma balthica*), 141
bamboo worm (red-banded bamboo worm *Axiothella rubrocincta*), 49
banded brittle star (*Ophionereis annulata*), 200
banded California marginella (*Volvarina taeniolata*), 109
banded chione (banded venus *Chione californiensis*), 147
banded tegula (banded turban *Tegula eiseni*), 83
banded tidepool fan (*Zonaria farlowii*), 227
banded turban (*Tegula eiseni*), 83

banded venus (*Chione californiensis*), 147
Bankia setacea (feathery shipworm), 157
bark semele (clipped semele *Semele decisa*), 145
barnacle-eating dorid (*Onchidoris bilamellata*), 115
barnacles, 162–65
baron delessert (winged fronds *Delesseria decipiens*), 249
basket cockle (Nuttall's cockle *Clinocardium nuttallii*), 137
bat star (*Asterina miniata*), 193
bat star worm (*Ophiodromus pugettensis*), 44
Batillaria attramentaria (mudflat snail *Batillaria cumingi*), 86
Batillaria cumingi (mudflat snail), 86
Batillaria zonalis (mudflat snail *Batillaria cumingi*), 86
bay cockle (rough-sided littleneck *Protothaca laciniata*), 149
bay ghost shrimp (*Neotrypaea californiensis*), 171
bay mussel (Pacific blue mussel *Mytilus trossulus*), 126
beach flea (California beach hopper *Megalorchestia californiana*), 168
beach sand anemone (moonglow anemone *Anthopleura artemisia*), 28
beaded anemone (red-beaded anemone *Urticina coriacea*), 31
beaded turban snail (banded turban *Tegula eiseni*), 83
beak thrower (proboscis worm *Glycera* sp.), 44
beaked piddock (*Netastoma rostratum*), 154
bean clam (*Donax gouldii*), 144
Bean Hollow State Beach, 266–67
bearded mussel (northern horsemussel *Modiolus modiolus*), 128
bent-nosed clam (bent-nose macoma *Macoma nasuta*), 142
bent-nose macoma (*Macoma nasuta*), 142
bifurcate mussel (branch-ribbed mussel *Septifer bifurcatus*), 127
big-neck clam (Pacific gaper *Tresus nuttallii*), 138
Bird Rock, 286
bivalves, 125–158
Bivalvia (class), 125–158

black abalone (*Haliotis cracherodii*), 71
black chiton (black Katy Chiton *Katharina tunicata*), 70
black-clawed crab (*Lophopanopeus bellus*), 184
black dog whelk (eastern mud snail *Ilyanassa obsoleta*, 106
black Katy Chiton (*Katharina tunicata*), 70
black limpet (*Lottia asmi*), 77
black pine (*Neorhodomela larix*), 250
black prickleback (*Xiphister atropurpureus*), 219
black sea hare (California black sea hare *Aplysia vaccaria*), 113
black tassel (*Pterosiphonia bipinnata*), 250
black-topped honeycomb worm (sand-castle worm *Phragmatopoma californica*), 50
black top-shell (black turban *Tegula funebralis*), 81
black turban (*Tegula funebralis*), 81
bladder wrack (Pacific rockweed *Fucus gardneri*), 235
bleached brunette (*Cryptosiphonia woodii*), 239
blind goby (*Typhlogobius californiensis*), 220
blister glassy-bubble (white bubble *Haminoea vesicula*), 111
blood star (Pacific blood star *Henricia leviuscula*), 195
bloodworm (*Euzonus* sp.), 48
bloodworm (proboscis worm *Glycera* sp.), 44
blue abalone (green abalone *Haliotis fulgens*), 72
blueband hermit (*Pagurus samuelis*), 173
blue-handed hermit crab (blueband hermit *Pagurus samuelis*), 173
blue-line chiton (*Tonicella undocaerulea*), 59
blue mussel (Pacific blue mussel *Mytilus trossulus*), 126
blue point oyster (Atlantic oyster *Crassostrea virginica*), 130
blue top shell (blue topsnail *Calliostoma ligatum*), 80
blue topsnail (*Calliostoma ligatum*), 80
blunt razor-clam (sickle jackknife clam *Solen sicarius*), 140
blunt-spined brittle star (flatspined brittle star *Ophiopteris papillosa*), 201
bodega clam (bodega tellin *Tellina bodegensis*), 143

bodega tellin (*Tellina bodegensis*), 143
Bolsa Chica State Beach, 282
Boltenia villosa (spiny-headed sea squirt), 203
boreal wentletrap (*Opalia borealis*), 88
boring softshell-clam (*Platyodon cancellatus*), 152
boring sponge (yellow boring sponge *Cliona celata*), 17
Botrylloides diegensis (orange chain tunicate), 213
Botryoglossum farlowianum. See fringed hidden rib (*Cryptopleura ruprechtiana*)
Botryoglossum ruprechtianum. See fringed hidden rib (*Cryptopleura ruprechtiana*)
Botula californiensis (California datemussel *Adula californiensis*), 128
Brachiopoda (phylum), 160
bracketed blenny (crescent gunnel *Pholis laeta*), 219
branched ribbed mussel (branch-ribbed mussel *Septifer bifurcatus*), 127
branched-spine bryozoan (*Flustrellidra corniculata*), 189
branch-ribbed mussel (*Septifer bifurcatus*), 127
bread crumb sponge (*Halichondria panicea*), 18
bristly tunicate (spiny-headed sea squirt *Boltenia villosa*), 203
brittle stars, 198–202
broad-disc sea star (bat star *Asterina miniata*), 193
broad-eared pecten (kelp scallop *Leptopecten latiauratus*), 132
broad iodine seaweed (*Prionitis lyallii*), 243
broad six-rayed sea star (six-rayed star *Leptasterias hexactis* species complex), 195
brooding anemone (proliferating anemone *Epiactis prolifera*), 32
brooding brittle star (dwarf brittle star *Amphipholis squamata*), 199
brown algae (phylum Phaeophyta), 222, 226–37
brown flatworm (tapered flatworm *Notocomplana acticola*), 38
brown intertidal spaghetti worm (*Eupolymnia heterobranchia*), 52
brown rock crab (spot-bellied rock crab *Cancer antennarius*), 183

brown-spotted nudibranch (ringed nudibranch *Diaulula sandiegensis*), 120
brown tegula (brown turban *Tegula brunnea*), 82
brown turban snail (brown turban *Tegula brunnea*), 82
brown turban (*Tegula brunnea*), 82
Bryozoa (phylum), 188–91
buckshot barnacle (little brown barnacle *Chthamalus dalli*), 162
Bulla gouldiana (California bubble), 110
bull kelp (*Nereocystis luetkeana*), 233
bullwhip kelp (bull kelp *Nereocystis luetkeana*), 233
buried sea anemone (moonglow anemone *Anthopleura artemisia*), 28
burrowing brittle star (long-armed brittle star *Amphiodia occidentalis*), 199
burrowing peanut worm (bushy-headed peanut worm *Themiste pyroides*), 55
bushy-backed nudibranch (*Dendronotus frondosus*), 121
bushy-backed sea slug (bushy-backed nudibranch *Dendronotus frondosus*), 121
bushy-headed peanut worm (*Themiste pyroides*), 55
butter clam (California butter clam *Saxidomus nuttalli*), 150
butter clam (*Saxidomus gigantea*). See under California butter clam (*Saxidomus nuttalli*)
butterfly chiton (regular chiton *Lepidozona regularis*), 63
butterfly crab (umbrella crab *Cryptolithodes sitchensis*), 174
butterfly scallop (San Diego scallop *Euvola diegensis*), 133
button shell (reticulate button snail *Trimusculus reticulatus*), 124
by-the-wind sailor (*Velella velella*), 24

Cabrillo National Monument, 286–87
Cabrillo's porcelain crab (*Petrolisthes cabrilloi*), 176
Cadlina luteomarginata (yellow-edged nudibranch), 117
Cadlina marginata (yellow-edged nudibranch *Cadlina luteomarginata*), 117
calcareous tube worm (*Serpula columbiana*), 54
California beach flea (California

beach hopper *Megalorchestia californiana*), 168
California beach hopper (*Megalorchestia californiana*), 168
California black sea hare (*Aplysia vaccaria*), 113
California brown cowry (chestnut cowrie *Zonaria spadicea*), 93
California brown sea hare (California sea hare *Aplysia californica*), 112
California bubble (*Bulla gouldiana*), 110
California butter clam (*Saxidomus nuttalli*), 150
California chiton (California spiny chiton *Nuttallina californica*), 64
California clingfish *Gobiesox rhessodon*. See under northern clingfish (*Gobiesox maeandricus*)
California cone (*Conus californicus*), 109
California cone shell (California cone *Conus californicus*), 109
California datemussel (*Adula californiensis*), 128
California fiddler (Mexican fiddler *Uca crenulata*), 187
California fireworm (*Pareurythoe californica*), 46
California ghost shrimp (bay ghost shrimp *Neotrypaea californiensis*), 171
California glass mya (California softshell-clam *Cryptomya californica*), 151
California grunion (*Leuresthes tenuis*), 217
California horn shell (California horn snail *Cerithidea californica*), 87
California hornsnail (California horn snail *Cerithidea californica*), 87
California horn snail (*Cerithidea californica*), 87
California ice cream cone worm (*Pectinaria californiensis*), 51
California jackknife clam (*Tagelus californianus*), 141
California limu (peppered spaghetti *Gracilaria pacifica*), 244
California lobster (California spiny lobster *Panulirus interruptus*), 171
California lucine (*Epilucina californica*), 134
California lyonsia (*Lyonsia californica*), 158

California mahogany-clam
  (*Nuttallia nuttallii*), 144
California marginella (banded
  California marginella
  Volvarina taeniolata), 109
California mussel (*Mytilus californianus*), 125
Californian chione (banded
  venus *Chione californiensis*),
  147
Californian irus venus (rock
  venus *Irusella lamellifera*),
  148
Californian tagelus (California
  jackknife clam *Tagelus californianus*), 141
California Nuttall chiton
  (California spiny chiton
  Nuttallina californica), 64
California oyster (Olympia oyster
  Ostrea conchaphila), 129
California pea-pod borer
  (California datemussel *Adula californiensis*), 128
California reversed chama
  (Pacific left-handed jewelbox
  Pseudochama exogyra), 135
California rock lobster
  (California spiny lobster
  Panulirus interruptus), 171
California sea cucumber
  (*Parastichopus californicus*),
  205
California sea hare (*Aplysia californica*), 112
California sea lace (*Microcladia californica*). See under delicate sea lace (*Microcladia coulteri*)
California sea pork (*Aplidium californicum*), 210
California softshell-clam
  (*Cryptomya californica*), 151
California spiny chiton
  (*Nuttallina californica*), 64
California spiny lobster
  (*Panulirus interruptus*), 171
California stichopus (California sea cucumber *Parastichopus californicus*), 205
California sunset clam (*Gari californica*), 145
California trivia (*Pusula californiana*), 93
California two spot octopus
  (two-spotted octopus *Octopus bimaculoides*), 159
California venus (banded venus
  Chione californiensis), 147
Callianassa californiensis (bay ghost shrimp *Neotrypaea californiensis*), 171
Calliarthron cheilosporioides.
  See coarse bead-coral alga

(*Calliarthron tuberculosum*)
Calliarthron regenerans (coarse
  bead-coral alga *Calliarthron tuberculosum*), 242
Calliarthron setchelliae. See
  coarse bead-coral alga
  (*Calliarthron tuberculosum*)
Calliarthron tuberculosum
  (coarse bead-coral alga), 242
Calliostoma costatum. See blue
  topsnail (*Calliostoma ligatum*)
Calliostoma ligatum (blue topsnail), 80
Cancer antennarius (spot-bellied rock crab), 183
Cancer magister (Dungeness
  Crab), 183
Cancer oregonensis (pygmy rock crab), 182
Cancer productus (red rock crab), 182
Caprella sp. (skeleton shrimp), 168
Carcinus maenas (green crab), 184
Cardium corbis (Nuttall's cockle *Clinocardium nuttallii*), 137
Cardium pedernalense. See
  Pacific eggcockle
  (*Laevicardium substriatum*)
carinated dove snail (carinate dovesnail *Alia carinata*), 102
carinate dove shell (carinate dovesnail *Alia carinata*), 102
carinate dovesnail (*Alia carinata*), 102
Carinella dinema. See six-lined ribbon worm (*Carinella sexlineata*)
Carinella rubra. See orange ribbon worm (*Tubulanus polymorphus*)
Carinella sexlineata (six-lined ribbon worm), 39
Carmel River State Beach, 270
Carpenter's carditid (little heart clam *Glans carpenteri*), 136
Carpinteria State Beach, 275
cats eyes (sea gooseberry
  (*Pleurobrachia bachei*), 27
Cayucos State Beach, 272
Cephalopoda (class), 159
Cephaloscyllium uter. See swell shark (*Cephaloscyllium ventriosum*)
Cephaloscyllium ventriosum
  (swell shark), 215
Ceramium codicola (staghorn fringe), 225
Ceratostoma foliatum (leafy hornmouth), 97
Ceratostoma nuttalli (Nuttall's hornmouth), 97

Cerebratulus montgomeryi (rose ribbon worm), 40
Cerithidea californica
  (California horn snail), 87
Chace's wentletrap (boreal wentletrap *Opalia borealis*), 88
Chaetopleura gemma (gem chiton), 66
chalk-lined dirona (frosted nudibranch *Dirona albolineata*), 121
Chama arcana (secret jewelbox), 135
Chama pellucida. See under
  secret jewelbox (*Chama arcana*)
channeled basket snail (giant western nassa *Nassarius fossatus*), 104
channeled dog whelk (giant western nassa *Nassarius fossatus*), 104
channeled purple (channelled dogwinkle *Nucella canaliculata*), 99
Channel Islands National Park, 276
channelled dogwinkle (*Nucella canaliculata*), 99
checked borer (boring softshell-clam *Platyodon cancellatus*), 152
checked softshell-clam (boring softshell-clam *Platyodon cancellatus*), 152
checkered littorine (checkered periwinkle *Littorina scutulata*), 85
checkered periwinkle (*Littorina scutulata*), 85
checkered thorn drupe (checkered unicorn *Acanthina paucilirata*), 96
checkered unicorn (*Acanthina paucilirata*), 96
chestnut cowrie (*Zonaria spadicea*), 93
chestnut cowry (chestnut cowrie *Zonaria spadicea*), 93
chevron amphiporus (two-spotted ribbon worm *Amphiporus bimaculatus*), 41
Chinaman's hat limpet (whitecap limpet *Acmaea mitra*), 75
Chione californiensis (banded venus), 147
Chione durhami. See banded venus (*Chione californiensis*)
Chione fluctifraga (smooth venus), 148
Chione succincta. See banded venus (*Chione californiensis*)
Chione taberi. See wavy venus (*Chione undatella*)

Chione undatella (wavy venus), 147
Chirolophis nugator (mosshead warbonnet), 218
chitons, 58–70
Chlorochytrium inclusum. See tangle weed (*Acrosiphonia coalita*)
Chlorophyta (phylum), 222–26
chocolate porcelain crab (*Petrolisthes manimaculis*), 177
Chondracanthus corymbiferus (Turkish towel), 245
Chone ecaudata (shell-binding colonial worm), 53
Chone minuta. See shell-binding colonial worm (*Chone ecaudata*)
Chordata (phylum), 214–20
Christmas anemone (painted anemone *Urticina crassicornis*), 33
Chrysaora fuscescens (sea nettle), 25
Chthamalus dalli (little brown barnacle), 162
chubby mya (boring softshell-clam *Platyodon cancellatus*), 152
circled dwarf triton (circled rocksnail *Ocenebra circumtexta*), 100
circled rock shell (circled rocksnail *Ocenebra circumtexta*), 100
circled rock snail (circled rocksnail *Ocenebra circumtexta*), 100
circled rocksnail (*Ocenebra circumtexta*), 100
Cirolana harfordi (scavenging isopod), 166
Cirripedia (class), 162–65
Cladophora sp. (sea moss), 224
clam hydroid (*Clytia bakeri*), 21
clamworm (pile worm *Nereis vexillosa*), 45
Clathromorphum parcum (stalked coralline disk), 240
Clavelina huntsmani (lightbulb tunicate), 210
clear jewel box (secret jewelbox *Chama arcana*), 135
Clinocardium nuttallii (Nuttall's cockle), 137
Cliona celata (yellow boring sponge), 17
clipped semele (*Semele decisa*), 145
cloudy bubble snail (California bubble *Bulla gouldiana*), 110
clubbed sea squirt (yellow lobed tunicate *Eudistoma ritteri*), 211

club-tipped anemone (strawberry anemone (*Corynactis californica*), 35
club tunicate (mushroom tunicate *Distaplia occidentalis*), 212
club tunicate (stalked compound tunicate *Distaplia smithi*), 212
clustering aggregate anemone (aggregating anemone *Anthopleura elegantissima*), 29
Clymenella rubrocincta (red-banded bamboo worm *Axiothella rubrocincta*), 49
Clytia bakeri (clam hydroid), 21
coarse bead-coral alga (*Calliarthron tuberculosum*), 242
coarse-tubed pink spaghetti worm (*Thelepus crispus*), 51
coat-of-mail shells. See chitons
Cockerell's dorid (*Laila cockerelli*), 116
Cockerell's nudibranch (Cockerell's dorid *Laila cockerelli*), 116
cockle (Nuttall's cockle *Clinocardium nuttallii*), 137
cockscomb prickleback (high cockscomb *Anoplarchus purpurescens*), 218
Codium fragile (sea staghorn), 225
Codium setchellii (spongy cushion), 226
Collisella asmi. See black limpet (*Lottia asmi*)
Collisella digitalis (ribbed limpet *Lottia digitalis*), 77
Collisella instabilis (unstable limpet *Lottia instabilis*), 79
Collisella limatula. See file limpet (*Lottia limatula*)
Collisella pelta (shield limpet *Lottia pelta*), 75
Collisella scabra. See rough limpet (*Lottia scabra*)
colonial sand tube worm (sandcastle worm *Phragmatopoma californica*), 50
Columbella carinata (carinate dovesnail *Alia carinata*), 102
Columbian amphissa (wrinkled amphissa *Amphissa columbiana*), 103
comb jellies, 27
comb jelly (sea gooseberry *Pleurobrachia bachei*), 27
commercial crab (Dungeness crab *Cancer magister*), 183
commercial oyster (Atlantic oyster *Crassostrea virginica*), 130

common acorn barnacle (acorn barnacle *Balanus glandula*), 163
common brown rockweed (Pacific rockweed *Fucus gardneri*), 235
common California venus (banded venus *Chione californiensis*), 147
common coastal shrimp (Sitka shrimp *Heptacarpus sitchensis*), 169
common coral seaweed (tidepool coralline alga *Corallina officinalis* var. *chilensis*), 241
common flatworm (tapered flatworm *Notocomplana acticola*), 38
common goose barnacle (pelagic goose barnacle *Lepas anatifera*), 165
common lampshell (*Terebratalia transversa*), 160
common Pacific brachiopod (common lampshell *Terebratalia transversa*), 160
common Pacific egg cockle (Pacific eggcockle *Laevicardium substriatum*), 137
common peanut worm (bushy-headed peanut worm *Themiste pyroides*), 55
common pickleweed (*Salicornia virginica*), 251, 252
common piddock (flat-tip piddock *Penitella penita*), 155
common rock crab (spot-bellied rock crab *Cancer antennarius*), 183
common sea cucumber (California sea cucumber *Parastichopus californicus*), 205
common sea star (purple star *Pisaster ochraceus*), 197
common Washington clam (California butter clam *Saxidomus nuttalli*), 150
conquina (bean clam *Donax gouldii*), 144
conspicuous chiton (*Stenoplax conspicua*), 66
Conus californicus (California cone), 109
Conus ravus (California cone *Conus californicus*), 109
convoluted sea fungus (*Petrospongium rugosum*), 227
Cooper's chiton (*Lepidozona cooperi*), 62
copper-haired sea mouse (*Aphrodita japonica*), 46

Corallina chilensis. See tidepool coralline alga (Corallina officinalis var. chilensis)
Corallina officinalis var. chilensis (tidepool coralline alga), 241
Corallina tuberculosa. See coarse bead-coral alga (Calliarthron tuberculosum)
corals, 27–28
corkscrew sea lettuce (Ulva taeniata), 223
Corynactis californica (strawberry anemone), 35
Coryphella fisheri. See three lined aeolid (Flabellina trilineata)
Coryphella sabulicola. See Spanish shawl (Flabellina iodinea)
Coryphella trilineata. See three lined aeolid (Flabellina trilineata)
Costaria costata (seersucker), 230
crabs, 172–86
Crangon stylirostris (smooth bay shrimp), 170
Crassadoma gigantea (giant rock scallop), 131
Crassostrea gigas (Pacific oyster), 130
Crassostrea virginica (Atlantic oyster), 130
Crepidula adunca (hooked slippersnail), 89
Crepidula cerithicola. See onyx slippersnail (Crepidula onyx)
Crepidula lirata. See onyx slippersnail (Crepidula onyx)
Crepidula nummaria (northern white slipper snail), 91
Crepidula onyx (onyx slippersnail), 90
Crepatella dorsata (Pacific half-slippersnail), 90
Crepipatella lingulata (Pacific half-slippersnail Crepipatella dorsata), 90
crescent gunnel (Pholis laeta), 219
crested blenny (high cockscomb Anoplarchus purpurescens), 218
crimson doris (red nudibranch Rostanga pulchra), 117
crisp leather (papillate seaweed/sea tar Mastocarpus papillatus), 246
Crucibulum spinosum (spiny cup-and-saucer), 91
crumb of bread sponge (bread crumb sponge Halichondria panicea), 18

cryptic kelp crab (Pugettia richii), 180
cryptic nudibranch (Doridella steinbergae), 113
cryptic spiny chiton (Nuttallina sp. Nov.). See under southern spiny chiton (Nuttallina scabra)
Cryptochiton stelleri (giant Pacific chiton), 69
Cryptolithodes sitchensis (umbrella crab), 174
Cryptomya californica (California softshell-clam), 151
Cryptopleura ruprechtiana (fringed hidden rib), 249
Cryptosiphonia grayana. See bleached brunette (Cryptosiphonia woodii)
Cryptosiphonia woodii (bleached brunette), 239
Crystal Cove State Park, 284
Ctenophora (phylum), 21, 27–35
Cumagloia andersonii (hairy seaweed), 239
Cuming's false cerith (mudflat snail Batillaria cumingi), 86
cup-and-saucer limpet (spiny cup-and-saucer Crucibulum spinosum), 91
cup corals, 27–28
Cyanea capillata (lion's mane), 26
Cyanoplax dentiens. See Gould's baby chiton (Lepidochiton dentiens)
Cyanoplax hartwegii. See Hartweg's chiton (Lepidochitona hartwegii)
Cylindrocarpus rugosus. See convoluted sea fungus (Petrospongium rugosum)
Cypraea spadicea. See chestnut cowrie (Zonaria spadicea)
Cystoseira osmundacea (giant bladder chain), 236

dahlia anemone (painted anemone Urticina crassicornis), 33
daisy brittle star (Ophiopholis aculeata), 198
dark-backed isopod (scavenging isopod Cirolana harfordi), 166
dark dwarf-turban (Homalopoma luridum), 84
dead man's fingers (sea sacs Halosaccion glandiforme), 221, 247
dead man's fingers (sea staghorn Codium fragile), 225
decorator crab (graceful decorator crab Oregonia gracilis), 178

Delesseria decipiens (winged fronds), 249
delicate sea lace (Microcladia coulteri), 248
Del Norte Coast Redwoods State Park, 257–58
Del Norte County, 257–58
Dendraster excentricus (eccentric sand dollar), 204
Dendrobeania lichenoides (sealichen bryozoan), 190
Dendrodoris fulva. See white spotted sea goddess (Doriopsilla albopunctata)
Dendronotus albus (white dendronotid), 120
Dendronotus arborescens (bushy-backed nudibranch Dendronotus frondosus), 121
Dendronotus frondosus (bushy-backed nudibranch), 121
Dendronotus venustus (bushy-backed nudibranch Dendronotus frondosus), 121
Dendrostomum petraeum. See bushy-headed peanut worm (Themiste pyroides)
Dendrostomum pyroides. See bushy-headed peanut worm (Themiste pyroides)
dense clumped laminarian (oar weed Laminaria sinclairii), 229
Dentrostoma patraeum. See bushy-headed peanut worm (Themiste pyroides)
Derbesia marina (sea pearls), 225
derby hat bryozoan (Eurystomella bilabiata), 188, 191
Dermasterias imbricata (leather star), 194
Desmarestia ligulata (flat acid kelp), 228
Diamphiodia occidentalis. See long-armed brittle star (Amphiodia occidentalis), 199
Diaulula sandiegensis (ringed nudibranch), 120
Dictyoneurum californicum (netted blade), 233
Diodora aspera (rough keyhole limpet), 73
Dirona albolineata (frosted nudibranch), 121
Dirona picta (painted nudibranch), 122
Discurria insessa (seaweed limpet), 80
Distaplia occidentalis (mushroom tunicate), 212
Distaplia smithi (stalked compound tunicate), 212

dogwinkles, 99–100
Donax californicus (wedge clam). See under bean clam (*Donax gouldii*)
Donax gouldi (bean clam *Donax gouldii*), 144
Donax gouldii (bean clam), 144
Doridella steinbergae (cryptic nudibranch), 113
Doriopsilla albopunctata (white spotted sea goddess), 119
Doriopsilla gemela (yellow-gilled sea goddess), 119
dotted pandora (punctate pandora *Pandora punctata*), 158
double pompom kelp (*Eisenia arborea*), 232
dulse (laver *Porphyra* sp.), 238
dunce-cap limpet (whitecap limpet *Acmaea mitra*), 75
Dungeness crab (*Cancer magister*), 183
dwarf brittle star (*Amphipholis squamata*), 199
dwarf rockweed (little rockweed *Pelvetiopsis limitata*), 235
dwarf teardrop crab (*Pelia tumida*), 181
dwarf triton (lurid rocksnail *Ocinebrina lurida*), 101

eastern mud nassa (eastern mud snail *Ilyanassa obsoleta*, 106
eastern mud snail (*Ilyanassa obsoleta*), 106
eastern mud whelk (eastern mud snail *Ilyanassa obsoleta*, 106
eastern oyster (Atlantic oyster *Crassostrea virginica*), 130
eastern oyster drill (Atlantic oyster drill *Urosalpinx cinerea*), 102
eccentric sand dollar (*Dendraster excentricus*), 204
Echinodermata (phylum), 192–206
Echinoidea (class), 202
echiuran worms, 56
Echiura (phylum), 56
edible cancer crab (Dungeness crab *Cancer magister*), 183
edible mussel (Pacific blue mussel *Mytilus trossulus*), 126
eel-grass (*Zostera* sp.), 253
egg cockle (Pacific eggcockle *Laevicardium substriatum*), 137
Egregia laevigata. See feather boa kelp (*Egregia menziesii*)
Egregia menziesii (feather boa kelp), 231
Egregia planifolia. See feather boa kelp (*Egregia menziesii*)

Eisenia arborea (double pompom kelp), 232
El Capitán State Beach, 274–75
elegant eolid (Spanish shawl *Flabellina iodinea*), 122
Elkhorn Sough, 268
emarginate dogwinkle (striped dogwinkle *Nucella emarginata*), 100
Emerita analoga (pacific mole crab), 178
Emplectonema gracile (green ribbon worm), 42
Emplectonema viride. See green ribbon worm (*Emplectonema gracile*)
Enchiridium punctatum (long speckled flatworm), 39
encrusted hairy chiton (Hind's Mopalia *Mopalia hindsii*), 69
encrusting bryozoan (kelp encrusting bryozoan *Membranipora membranacea*), 190
encrusting coral (encrusting coralline algae *Lithothamnion* sp), 240
encrusting coralline algae (*Lithothamnion* sp), 240
encrusting sponge (purple encrusting sponge *Haliclona permollis*), 19
Endocladia muricata (nail brush seaweed), 242
Enteromorpha sp. (tubeweed), 222
Entodesma navicula (rock entodesma), 157
Entodesma saxicola (rock entodesma *Entodesma navicula*), 157
Epiactis prolifera (proliferating anemone), 32
Epialtus nuttallii. See globose kelp crab (*Taliepus nuttalli*)
Epigeichthys atropurpureus (black prickleback *Xiphister atropurpureus*), 219
Epilucina californica (California lucine), 134
Epitonium tinctum (tinted wentletrap *Nitidiscala tincta*), 87
Erato vitellina. See appleseed erato (*Hespererato vitellina*)
eroded periwinkle (flat-bottomed periwinkle *Littorina keenae*), 85
Erythrophyllum delesserioides (red sea-leaf), 243
Esmark's brittle star (smooth brittle star *Ophioplocus esmarki*), 201
Esmark's serpent star (smooth brittle star *Ophioplocus esmarki*), 201
Eudistoma ritteri (yellow lobed tunicate), 211
Eupentacta quinquesemita (stiff-footed sea cucumber), 206
Eupolymnia heterobranchia (brown intertidal spaghetti worm), 52
European green crab (green crab *Carcinus maenas*), 184
European pickleweed (*S. europaea*). See under common pickleweed (*Salicornia virginica*)
European shore crab (green crab *Carcinus maenas*), 184
Eurystomella bilabiata (derby hat bryozoan), 188, 191
Euspira lewisii (Lewis's moonsnail), 94
Euvola diegensis (San Diego scallop), 133
Euzonus sp. (bloodworm), 48

false-cerith snail (mudflat snail *Batillaria cumingi*), 86
false mya (California softshell-clam *Cryptomya californica*), 151
false Pacific jingle shell (green false-jingle *Pododesmus macroschisma*), 134
fan horsemussel (straight horse-mussel *Modiolus rectus*), 129
fan-shaped horse mussel (straight horsemussel *Modiolus rectus*), 129
fan shell (Pacific calico scallop *Argopecten ventricosus*), 132
fat horse mussel (straight horse-mussel *Modiolus rectus*), 129
fat innkeeper worm (*Urechis caupo*), 56
fat nassa (western fat dog nassa *Nassarius perpinguis*), 105
feather boa kelp (*Egregia menziesii*), 231
feathery shipworm (*Bankia setacea*), 157
felty fingers (sea staghorn *Codium fragile*), 225
festive murex (*Pteropurpura festiva*), 98
festive rock shell (festive murex *Pteropurpura festiva*), 98
fiddler crab (Mexican fiddler *Uca crenulata*), 185
file limpet (*Lottia limatula*), 78
file shell (Hemphill fileclam *Limaria hemphilli*), 131
fine-tubed pink spaghetti worm (*Thelepus setosus*). See under coarse-tubed pink spaghetti worm (*Thelepus crispus*)

fingered limpet (ribbed limpet *Lottia digitalis*), 77
finger limpet (ribbed limpet *Lottia digitalis*), 77
First Peoples. *See* Native people
fishes, 214–20
Fissurella volcano (volcano keyhole limpet), 73
Fissurellidea bimaculata (two-spot keyhole limpet), 74
Fitzgerald, James V., Marine Reserve, 266
five-ribbed kelp (seersucker *Costaria costata*), 230
Flabellina iodinea (Spanish shawl), 122
Flabellina trilineata (three lined aeolid), 123
Flabellinopsis iodinea. *See* Spanish shawl (*Flabellina iodinea*)
flat acid kelp (*Desmarestia ligulata*), 228
flat-bottomed periwinkle (*Littorina keenae*), 85
flathead clingfish (northern clingfish *Gobiesox maeandricus*), 216
flat hoofsnail (*Hipponix cranioides*), 89
flat porcelain crab (*Petrolisthes cinctipes*), 176
flat-spined brittle star (*Ophiopteris papillosa*), 201
flattened acid kelp (flat acid kelp *Desmarestia ligulata*), 228
flat-tipped wire alga (*Mastocarpus jardinii*), 245
flat-tip piddock (*Penitella penita*), 155
flattop crab (*Petrolisthes eriomerus*), 161, 175
flat-topped crab (flattop crab *Petrolisthes eriomerus*), 175
flatworms, 37–39
flowering peanut worm (bushy-headed peanut worm *Themiste pyroides*), 55
flowering plants, 251–53
Flustrella cervicornis (branched-spine bryozoan *Flustrellidra corniculata*), 189
Flustrellidra corniculata (branched-spine bryozoan), 189
fluted bryozoan (*Hippodiplosia insculpta*), 191
foolish mussel (Pacific blue mussel *Mytilus trossulus*), 126
forked kelp (double pompom kelp *Eisenia arborea*), 232
forty-ribbed cockle (spiny cockle *Trachycardium quadragenarium*), 138

fragile scaleworm (*Arctonoe fragilis*), 47
frilled anemone (plumose anemone *Metridium senile*), 34
frilled California venus (wavy venus *Chione undatella*), 147
frilled commensal scaleworm (fragile scaleworm *Arctonoe fragilis*), 47
frilled dogwinkle (*Nucella lamellosa*), 99
frilled hoof shell (Pacific left-handed jewelbox *Pseudochama exogyra*), 135
frilled hoof shell (secret jewelbox *Chama arcana*), 135
frilled venus (wavy venus *Chione undatella*), 147
frilled whelk (frilled dogwinkle *Nucella lamellosa*), 99
fringed hidden rib (*Cryptopleura ruprechtiana*), 249
frosted nudibranch (*Dirona albolineata*), 121
Fucus distichus. *See* Pacific rockweed (*Fucus gardneri*)
Fucus elminthoides. *See* rubber threads (*Nemalion helminthoides*)
Fucus evanescens. *See* Pacific rockweed (*Fucus gardneri*)
Fucus fastigiatus. *See* spindle-shaped rockweed (*Pelvetia fastigiata*)
Fucus furcatus. *See* Pacific rockweed (*Fucus gardneri*)
Fucus gardneri (Pacific rockweed), 235
Fucus ligatulus. *See* flat acid kelp (*Desmarestia ligulata*)
furry crab (*Hapalogaster cavicauda*), 174
Fusinus luteopictus (painted spindle), 106
fuzzy crab (furry crab *Hapalogaster cavicauda*), 174

Gadinia reticulata. *See* reticulate button snail *Trimusculus reticulatus*)
Gallic saxicave (Arctic hiatella *Hiatella arctica*), 154
gaper pea crab (*Pinnixa littoralis*), 186
Gari californica (California sunset clam), 145
garlic star (leather star *Dermasterias imbricata*), 194
Garrapata State Beach, 271
Gastropoda (class), 71–111
gastropods, 71–111
gem chiton (*Chaetopleura gemma*), 66

gem murex (*Maxwellia gemma*), 98
gem rock shell (gem murex *Maxwellia gemma*), 98
Geukensia demissa (ribbed mussel), 127
ghost shrimp (bay ghost shrimp *Neotrypaea californiensis*), 171
giant bladder chain (*Cystoseira osmundacea*), 236
giant chiton (giant Pacific chiton *Cryptochiton stelleri*), 69
giant green anemone (*Anthopleura xanthogrammica*), 30
giant horse mussel (straight horse-mussel *Modiolus rectus*), 129
giant kelp (giant perennial kelp *Macrocystis pyrifera*), 234
giant kelp (small perennial kelp *Macrocystis integrifolia*), 234
giant key hole limpet (*Megathura crenulata*), 74
giant owl limpet (*Lottia gigantea*), 79
giant Pacific chiton (*Cryptochiton stelleri*), 69
giant Pacific cockle (spiny cockle *Trachycardium quadragenarium*), 138
giant Pacific oyster (Pacific oyster *Crassostrea gigas*), 130
giant perennial kelp (*Macrocystis pyrifera*), 234
giant pink star (*Pisaster brevispinus*), 196
giant red chiton (giant Pacific chiton *Cryptochiton stelleri*), 69
giant red urchin (red sea urchin *Strongylocentrotus franciscanus*), 203
giant rock scallop (*Crassadoma gigantea*), 131
giant sea cucumber (California sea cucumber *Parastichopus californicus*), 205
giant sea star (giant spined star *Pisaster giganteus*), 196
giant spined star (*Pisaster giganteus*), 196
giant western nassa (*Nassarius fossatus*), 104
Gigartina agardhii. *See* flat-tipped wire alga (*Mastocarpus jardinii*)
Gigartina corymbifera. *See* Turkish towel (*Chondracanthus corymbiferus*)
Gigartina cristata. *See* papillate seaweed/sea tar (*Mastocarpus papillatus*)

Gigartina papillata. See papillate seaweed/sea tar (*Mastocarpus papillatus*)
gilded tegula (gilded turban *Tegula aureotincta*), 84
gilded turban (*Tegula aureotincta*), 84
Glans carpenteri (little heart clam), 136
Glans minuscula. See little heart clam (*Glans carpenteri*)
Glans subquadrata. See little heart clam (*Glans carpenteri*)
globose kelp crab (*Taliepus nuttalli*), 180
Glycera sp. (proboscis worm), 44
Gobiesox maeandricus (northern clingfish), 216
gooeyduck (Pacific geoduck *Panope abrupta*), 153
goose barnacle (*Pollicipes polymerus*), 165
gooseneck barnacle (goose barnacle *Pollicipes polymerus*), 165
Gould beanclam (bean clam *Donax gouldii*), 144
Gould's Baby Chiton (*Lepidochiton dentiens*), 60
Gould's bubble (California bubble *Bulla gouldiana*), 110
Gould's bubble shell (California bubble *Bulla gouldiana*), 110
graceful decorator crab (*Oregonia gracilis*), 178
graceful kelp crab (*Pugettia gracilis*), 179
graceful rock crab (graceful kelp crab *Pugettia gracilis*), 179
Gracilaria confervoides. See peppered spaghetti (*Gracilaria pacifica*)
Gracilaria pacifica (peppered spaghetti), 244
Gracilaria verrucosa. See peppered spaghetti (*Gracilaria pacifica*)
grainyhand hermit (*Pagurus granosimanus*), 172
granular claw crab (*Oedignathus inermis*), 175
granular hermit crab (grainyhand hermit *Pagurus granosimanus*), 172
gray littorine (flat-bottomed periwinkle *Littorina keenae*), 85
gray periwinkle (flat-bottomed periwinkle *Littorina keenae*), 85
gray snakeskin-snail (*Ophiodermella inermis*), 110
great keyhole limpet (giant keyhole limpet *Megathura crenulata*), 74

great Washington clam (Pacific gaper *Tresus nuttallii*), 138
green abalone (*Haliotis fulgens*), 72
green algae (phylum Chlorophyta), 222–26
green and yellow ribbon worm (green ribbon worm *Emplectonema gracile*), 42
green anemone (giant green anemone (*Anthopleura xanthogrammica*), 30
green bubble (*Haminoea virescens*), 111
green burrowing anemone (moonglow anemone *Anthopleura artemisia*), 28
green crab (*Carcinus maenas*), 184
green false-jingle (*Pododesmus macroschisma*), 134
green isopod (Vosnesensky's isopod *Idotea wosnesenskii*), 167
green paper bubble (green bubble *Haminoea virescens*), 111
green ribbon worm (*Emplectonema gracile*), 42
green rope (tangle weed *Acrosiphonia coalita* gametophyte stage), 224
green sea grape (sea pearls *Derbesia marina*), 225
green sea velvet (sea staghorn *Codium fragile*), 225
grunion (California grunion *Leuresthes tenuis*), 217
gumboot chiton (giant Pacific chiton *Cryptochiton stelleri*), 69
Gymnomorpha (subclass), 124

habitats, intertidal, 10–12
hairy cancer crab (pygmy rock crab *Cancer oregonensis*), 182
hairy chiton (*Mopalia ciliata*), 67
hairy crab (retiring southerner *Pilumnus spinohirsutus*), 185
hairy gilled worm (coarse-tubed pink spaghetti worm *Thelepus crispus*), 51
hairy-headed terebellid worm (coarse-tubed pink spaghetti worm *Thelepus crispus*), 51
hairy hermit (*Pagurus hirsutiusculus*), 173
hairy lithodid crab (furry crab *Hapalogaster cavicauda*), 174
hairy sea squirt (spiny-headed sea squirt *Boltenia villosa*), 203
hairy seaweed (*Cumagloia andersonii*), 239
hairy under-rock crab (retiring southerner *Pilumnus spinohirsutus*), 185
half-fringed sea fern (*Neoptilota densa*), 248
half-pitted mitre (Ida's mitre *Mitra idae*), 108
half-slipper shell (Pacific half-slippersnail *Crepipatella dorsata*), 90
half-slipper snail (Pacific half-slippersnail *Crepipatella dorsata*), 90
Halichondria panicea (bread crumb sponge), 18
Haliclona permollis (purple encrusting sponge), 19
Halicystis ovalis. See sea pearls (*Derbesia marina*)
Haliotis cracherodii (black abalone), 71
Haliotis fulgens (green abalone) 72
Haliotis rufescens (red abalone), 72
Haliplanella lineata. See striped anemone (*Haliplanella luciae*)
Haliplanella luciae (striped anemone), 32
Halosaccion glandiforme (sea sacs), 247
Haminoea cymbiformis (green bubble *Haminoea virescens*), 111
Haminoea strongi (green bubble *Haminoea virescens*), 111
Haminoea vesicula (white bubble), 111
Haminoea virescens (green bubble), 111
Hapalogaster cavicauda (furry crab), 174
hardshell clam (Pacific littleneck *Protothaca staminea*), 150
Harford's greedy isopod (scavenging isopod *Cirolana harfordi*), 166
Hartweg's chiton (*Lepidochitona hartwegii*), 61
heart cockle (Nuttall's cockle *Clinocardium nuttallii*), 137
Heath's chiton (*Stenoplax heathiana*), 67
Hedophyllum sessile (sea cabbage), 229
helmet crab (*Telmessus cheiragonus*), 181
Hemigrapsus nudus (purple shore crab), 186
Hemphill fileclam (*Limaria hemphilli*), 131
Hemphill's dish clam (Hemphill surfclam *Mactromeris hemphillii*), 139

Hemphill's lima (Hemphill fileclam (*Limaria hemphilli*), 131
Hemphill's surfclam (Hemphill surfclam *Mactromeris hemphillii*), 139
Hemphill surfclam (*Mactromeris hemphillii*), 139
Henricia leviuscula (Pacific blood star), 195
Heptacarpus carinatus (small-eyed shrimp), 169
Heptacarpus pictus. See Sitka shrimp (*Heptacarpus sitchensis*)
Heptacarpus sitchensis (Sitka shrimp), 169
Hermissenda crassicornis (opalescent nudibranch), 124
hermissenda nudibranch (opalescent nudibranch *Hermissenda crassicornis*), 124
Hespererato vitellina (appleseed erato), 92
Hiatella arctica (Arctic hiatella), 154
Hiatella gallicana (Arctic hiatella *Hiatella arctica*), 154
hidden rib (fringed hidden rib *Cryptopleura ruprechtiana*), 249
high cockscomb (*Anoplarchus purpurescens*), 214, 218
Hind's mopalia (*Mopalia hindsii*), 69
Hinnites giganteus (giant rock scallop *Crassadoma gigantea*), 131
Hinnites multirugosus (giant rock scallop *Crassadoma gigantea*), 131
Hippodiplosia insculpta (fluted bryozoan), 191
Hippolysmata californica. See red rock shrimp (*Lysmata californica*)
Hipponix antiquatus (flat hoofsnail *Hipponix cranioides*), 89
Hipponix cranioides (flat hoofsnail), 89
holdfast brittle star (dwarf brittle star *Amphipholis squamata*), 199
Holopagurus pilosus. See moonsnail hermit (*Isocheles pilosus*)
Holothuroidea (class), 204–6
Homalopoma carpenteri. See dark dwarf-turban (*Homalopoma luridum*)
Homalopoma luridum (dark dwarf-turban), 84
hooked slippersnail (*Crepidula adunca*), 89

Hopkinsia rosacea (Hopkin's rose), 114
Hopkin's rose (*Hopkinsia rosacea*), 114
hornmouths, 97
horse clam (Pacific gaper *Tresus nuttallii*), 138
horse crab (helmet crab *Telmessus cheiragonus*), 181
horse mussel (northern horse-mussel *Modiolus modiolus*), 128
horse's hoofsnail (flat hoofsnail *Hipponix cranioides*), 89
Humboldt County, 258–59
Hyalina californica. See banded California marginella (*Volvarina taeniolata*)
hydroids, 21–23
Hydrozoa (class), 21

ice cream cone worm (California ice cream cone worm *Pectinaria californiensis*), 51
Ida's mitre (*Mitra idae*), 108
Idotea wosnesenskii (Vosnesensky's isopod), 167
Ilyanassa obsoleta (eastern mud snail), 106
Ilyanassa obsoletus. See eastern mud snail (*Ilyanassa obsoleta*)
indented macoma (*Macoma indentata*), 142
innkeeper worm (fat innkeeper worm *Urechis caupo*), 56
intertidal mite (red velvet mite *Neomolgus littoralis*), 187
intertidal sites, 255–87; north coast, 256–64; central coast, 265–76; south coast, 277–87
iodine seaweed (broad iodine seaweed *Prionitis lyallii*), 243
Iridaea cordata. See iridescent seaweed (*Mazzaella splendens*)
iridescent seaweed (*Mazzaella splendens*), 247
Iridophycus splendens. See iridescent seaweed (*Mazzaella splendens*)
Irusella lamellifera (rock venus), 148
Irus lamellifer. See rock venus (*Irusella lamellifera*)
Ischnochiton californiensis. See pectinate lepidozona (*Lepidozona pectinulata*)
Ischnochiton conspicuus. See conspicuous chiton (*Stenoplax conspicua*)
Ischnochiton cooperi. See Cooper's chiton (*Lepidozona cooperi*)

Ischnochiton heathiana. See Heath's chiton (*Stenoplax heathiana*)
Ischnochiton interstinctus. See smooth lepidozona (*Lepidozona intersticta*)
Ischnochiton marmortus. See gem chiton (*Chaetopleura gemma*)
Ischnochiton radians. See smooth lepidozona (*Lepidozona intersticta*)
Ischnochiton regularis (regular chiton *Lepidozona regularis*), 63
Isocheles pilosus (moonsnail hermit), 172
isopods, 166–67

James V. Fitzgerald Marine Reserve, 266
Japanese littleneck clam (Japanese littleneck *Venerupis philippinarum*), 149
Japanese littleneck (*Venerupis philippinarum*), 149
Japanese oyster (Pacific oyster *Crassostrea gigas*), 130
japweed (sargassum *Sargassum muticum*), 237
Jasminiera ecaudata. See shell-binding colonial worm (*Chone ecaudata*)
Jaton festivus. See festive murex (*Pteropurpura festiva*)
jellies, 20, 23–27
jingle shell (green false-jingle *Pododesmus macroschisma*), 134
jingle shell (pearly jingle *Anomia peruviana*), 133
jointworm (red-banded bamboo worm *Axiothella rubrocincta*), 49
Joseph's coat amphissa (variegated amphissa *Amphissa versicolor*), 103

Kaburakia excelsa (large leaf worm), 37
Katharina tunicata (black Katy Chiton), 70
keeled dove shell (carinate dovesnail *Alia carinata*), 102
Keep's Chiton (*Lepidochitona keepiana*), 60
Kelletia kelletii (Kellet's whelk), 104
Kelletia kelleti (Kellet's whelk *Kelletia kelletii*), 104
Kellet's whelk (*Kelletia kelletii*), 104
Kellia comandorica. See kellyclam (*Kellia suborbicularis*)

299

Kellia laperousii. See kellyclam (*Kellia suborbicularis*)
Kellia suborbicularis (kellyclam), 136
kellyclam (*Kellia suborbicularis*), 136
kelp, 228, 230, 231, 232
kelp crab (graceful kelp crab *Pugettia gracilis*), 179
kelp crab (shield-backed kelp crab *Pugettia producta*), 179
kelp encrusting bryozoan (*Membranipora membranacea*), 190
kelp lace (kelp encrusting bryozoan *Membranipora membranacea*), 190
kelp limpet (seaweed limpet *Discurria insessa*), 80
kelp scallop (*Leptopecten latiauratus*), 132
kelp-weed scallop (kelp scallop *Leptopecten latiauratus*), 132
keyhole limpet (volcano keyhole limpet *Fissurella volcano*), 73
king clam (Pacific geoduck *Panope abrupta*), 153
knobby sea star (giant spined star *Pisaster giganteus*), 196
knobby starfish (giant spined star *Pisaster giganteus*), 196

lacy-crust bryozoan (kelp encrusting bryozoan *Membranipora membranacea*), 190
Laevicardium substriatum (Pacific eggcockle), 137
Laila cockerelli (Cockerell's dorid), 116
lamellar venus (rock venus *Irusella lamellifera*), 148
Lamellaria diegoensis (San Diego lamellaria), 92
Lamellaria stearnsii. See San Diego lamellaria (*Lamellaria diegoensis*)
Laminaria sessilis. See sea cabbage (*Hedophyllum sessile*)
Laminaria setchellii (split kelp), 228
Laminaria sinclairii (oar weed), 229
lampshell (common lampshell *Terebratalia transversa*), 160
Lanice heterobranchia. See brown intertidal spaghetti worm (*Eupolymnia heterobranchia*)
large red cucumber (California sea cucumber *Parastichopus californicus*), 205
large leaf worm (*Kaburakia excelsa*), 37

laver (*Porphyra* sp.), 238
leaf barnacle (goose barnacle *Pollicipes polymerus*), 165
leaf worm (large leaf worm *Kaburakia excelsa*), 37
leafy hornmouth (*Ceratostoma foliatum*), 97
lean nassa (western lean nassa *Nassarius mendicus*), 105
leather chiton (black Katy Chiton *Katharina tunicata*), 70
leather star (*Dermasterias imbricata*), 194
leathery anemone (red-beaded anemone *Urticina coriacea*), 31
Leathesia difformis (sea cauliflower), 226
left-handed jewel box (Pacific left-handed jewelbox *Pseudochama exogyra*), 135
Leo Carrillo State Beach, 278
leopard nudibranch (ringed nudibranch *Diaulula sandiegensis*), 120
Lepas anatifera (pelagic goose barnacle), 165
Lepidochitona hartwegii (Hartweg's chiton), 61
Lepidochitona keepiana (Keep's chiton), 60
Lepidochiton dentiens (Gould's baby chiton), 60
Lepidochiton hartwegii. See Hartweg's chiton (*Lepidochitona hartwegii*)
Lepidozona californiensis. See pectinate lepidozona (*Lepidozona pectinulata*)
Lepidozona cooperi (Cooper's chiton), 62
Lepidozona intersticta (smooth lepidozona), 62
Lepidozona mertensii (Merten's chiton), 61
Lepidozona pectinulata (pectinate lepidozona), 63
Lepidozona regularis (regular chiton), 63
Leptasterias hexactis species complex (six-rayed star), 195
Leptopecten latiauratus (kelp scallop), 132
Leptoplana acticola. See tapered flatworm (*Notocomplana acticola*)
Leptothyra carpenteri. See dark dwarf-turban (*Homalopoma luridum*)
lesser California sea cucumber (warty sea cucumber *Parastichopus parvimensis*), 206

Lessonia sinclairii. See oar weed (*Laminaria sinclairii*)
Leuresthes tenuis (California grunion), 217
Lewis's moonsnail (*Euspira lewisii*), 94
lightbulb ascidian (lightbulb tunicate *Clavelina huntsmani*), 207, 210
lightbulb tunicate (*Clavelina huntsmani*), 207, 210
Ligia occidentalis (western sea roach), 167
Lima hemphilli. See Hemphill fileclam (*Limaria hemphilli*)
Limaria hemphilli (Hemphill fileclam), 131
limpets, 73–80
lined chiton (*Tonicella lineata*), 58. See also Loki's chiton (*Tonicella lokii*); blue-line chiton (*Tonicella undocaerulea*)
lined red chiton (lined chiton *Tonicella lineata*), 58
lined ribbon worm (six-lined ribbon worm *Carinella sexlineata*), 39
Lineus pictifrons (velvet ribbon worm), 41
link confetti (tubeweed *Enteromorpha* sp.), 222
lion's mane (*Cyanea capillata*), 26
Liparis florae (tidepool snailfish), 217
Lithothamnion sp. (encrusting coralline algae), 240
Lithothamnium parcum. See stalked coralline disk (*Clathromorphum parcum*)
Lithothamnium sp. See encrusting coralline algae (*Lithothamnion* sp.)
Lithothrix aspergillum (stone hair), 241
little brown barnacle (*Chthamalus dalli*), 162
little coffee-bean (California trivia *Pusula californiana*), 93
little coffee bean shell (California trivia *Pusula californiana*), 93
Little Corona Del Mar Beach, 283
little egg cockle (Pacific eggcockle *Laevicardium substriatum*), 137
little gaper (Arctic hiatella *Hiatella arctica*), 154
little heart clam (*Glans carpenteri*), 136
little olive (baetic olive *Olivella baetica*), 107
little rockweed (*Pelvetiopsis limitata*), 235

little striped barnacle (*Balanus amphitrite*). See under white-ribbed red barnacle (*Megabalanus californicus*)
Littorina keenae (flat-bottomed periwinkle), 85
Littorina planaxis. See flat-bottomed periwinkle (*Littorina keenae*)
Littorina scutulata (checkered periwinkle), 85
Loki's chiton (*Tonicellalokii*), 59
long arm brittle star (long-armed brittle star *Amphiodia occidentalis*), 199
long-armed brittle star (*Amphiodia occidentalis*), 199
long-horned nudibranch (opalescent nudibranch *Hermissenda crassicornis*), 124
long speckled flatworm (*Enchiridium punctatum*), 39
long-stalked sea squirt (*Styela montereyensis*), 209
Lophopanopeus bellus (black-clawed crab), 184
Los Angeles County, 278–281
Lottia asmi (black limpet), 77
Lottia digitalis (ribbed limpet), 77
Lottia gigantea (giant owl limpet), 79
Lottia instabilis (unstable limpet), 79
Lottia limatula (file limpet), 78
Lottia pelta (shield limpet), 75
Lottia scabra (rough limpet), 78
Lover's Point, 268–69
Lucina californica. See California lucine (*Epilucina californica*)
lugworm (Pacific lugworm *Abarenicola pacifica*), 49
Lunatia lewisi (Lewis's moonsnail *Euspira lewisii*), 94
lurid rock snail (lurid rocksnail *Ocinebrina lurida*), 101
lurid rocksnail (*Ocinebrina lurida*), 101
Lyall's iodine seaweed (broad iodine seaweed *Prionitis lyallii*), 243
Lyonsia californica (California lyonsia), 158
Lysmata californica (red rock shrimp), 170

Macclintockia scabra. See rough limpet (*Lottia scabra*)
MacKerricher State Park, 260
Macoma balthica (Baltic macoma), 141
Macoma inconspicua (Baltic macoma *Macoma balthica*), 141
Macoma indentata (indented macoma), 142
Macoma nasuta (bent-nose macoma), 142
Macoma rickettsi. See indented macoma (*Macoma indentata*)
Macoma secta (white-sand macoma), 143
Macrocystis integrifolia (small perennial kelp), 234
Macrocystis pyrifera (giant perennial kelp), 234
Mactromeris hemphillii (Hemphill surfclam), 139
Malacostraca (class), 166–86
maned nudibranch (shag-rug nudibranch *Aeolidia papillosa*), 123
Manila clam (Japanese littleneck *Venerupis philippinarum*), 149
many-ribbed hydromedusa (water jelly *Aequorea* sp.), 23
many-ribbed jellyfish (water jelly *Aequorea* sp.), 23
Marginella californica. See banded California marginella (*Volvarina taeniolata*)
Marin County, 263–64
marine worms, 36–56
masked limpet (mask limpet *Tectura persona*), 76
mask limpet (*Tectura persona*), 76
Mastocarpus corymbiferus. See Turkish towel (*Chondracanthus corymbiferus*)
Mastocarpus jardinii (flat-tipped wire alga), 245
Mastocarpus papillatus (papillate seaweed/sea tar), 246
Maxwellia gemma (gem murex), 98
Maxwell's gem shell (gem murex *Maxwellia gemma*), 98
Mazzaella splendens (iridescent seaweed), 81
Mediterranean mussel (*Mytilus galloprovincialis*), 126
Megabalanus californicus (white-ribbed red barnacle), 164
Megalorchestia californiana (California beach hopper), 168
Megastraea undosa (wavy turban), 81
Megatebennus bimaculatus. See two-spot keyhole limpet (*Fissurellidea bimaculata*)
Megathura crenulata (giant keyhole limpet), 74

Membranipora membranacea (kelp encrusting bryozoan), 190
Mendocino County, 259–61
Mendocino Headlands State Park, 261
Merten's chiton (*Lepidozona mertensii*), 61
Metridium senile (plumose anemone), 34
Mexican fiddler (*Uca crenulata*), 187
Microcladia californica (California sea lace). See under delicate sea lace (*Microcladia coulteri*)
Microcladia coulteri (delicate sea lace), 248
Mitella polymerus. See goose barnacle (*Pollicipes polymerus*)
Mitra catalinae (Ida's mitre *Mitra idae*), 108
Mitra diegensis (Ida's mitre *Mitra idae*), 108
Mitra idae (Ida's mitre), 108
Mitra montereyi (Ida's mitre *Mitra idae*), 108
Mitrella carinata (carinate dovesnail *Alia carinata*), 102
Modiola plicatula. See ribbed mussel (*Geukensia demissa*)
Modiola semicostata. See ribbed mussel (*Geukensia demissa*)
Modiolus modiolus (northern horsemussel), 128
Modiolus rectus (straight horsemussel), 129
Mollusca (phylum), 57–111
mollusks, 57–111
Montaña de Oro State Park, 272–73
Monterey dorid (*Archidoris montereyensis*), 118
Monterey sea lemon (Monterey dorid *Archidoris montereyensis*), 118
Monterey stalked tunicate (long-stalked sea squirt *Styela montereyensis*), 209
moonglow anemone (*Anthopleura artemisia*), 28
moon jelly (*Aurelia labiata*), 26
moon jellyfish (moon jelly (*Aurelia labiata*), 26
moonsnail hermit (*Isocheles pilosus*), 172
Mopalia ciliata (hairy chiton), 67
Mopalia hindsii (Hind's mopalia), 69
Mopalia lignosa (woody chiton), 68

Mopalia muscosa (mossy chiton), 68
Mopalia wosnessenskii (hairy chiton *Mopalia ciliata*), 67
moss animals, 188–91
mosshead prickleback (mosshead warbonnet *Chirolophis nugator*), 218
mosshead warbonnet (*Chirolophis nugator*), 218
mossy chiton (*Mopalia muscosa*), 68
mottled anemone (painted anemone *Urticina crassicornis*), 33
mucus sponge (*Plocamia karykina*), 16, 19
mud clam (softshell-clam *Mya arenaria*), 152
mud dog whelk (eastern mud snail *Ilyanassa obsoleta*, 106
mud flat octopus (two-spotted octopus *Octopus bimaculoides*), 159
mudflat snail (*Batillaria cumingi*), 86
mud nemertean (purple-backed ribbon worm *Paranemertes peregrina*), 43
multi-lobed tunicate (*Ritterella aequalisiphonis*), 213
mushroom tunicate (*Distaplia occidentalis*), 212
mussel worm (pile worm *Nereis vexillosa*), 45
Mya arenaria (softshell-clam), 152
Mytilus californianus (California mussel), 125
Mytilus demissus. See ribbed mussel (*Geukensia demissa*)
Mytilus edulis diegensis. See Mediterranean mussel (*Mytilus galloprovincialis*)
Mytilus edulis. See Pacific blue mussel (*Mytilus trossulus*)
Mytilus galloprovincialis (Mediterranean mussel), 126
Mytilus trossulus (Pacific blue mussel), 126

nail brush seaweed (*Endocladia muricata*), 242
Nanaimo dorid (rufus tipped nudibranch *Acanthodoris nanaimoensis*), 114
Nassarius cooperi (western lean nassa *Nassarius mendicus*), 105
Nassarius fossatus (giant western nassa), 104
Nassarius mendicus (western lean nassa), 105
Naśsarius obsoletus. See eastern mud snail (*Ilyanassa obsoleta*)
Nassarius perpinguis (western fat dog nassa), 105
native Pacific oyster (Olympia oyster *Ostrea conchaphila*), 129
Native people, 30, 69, 70, 71, 74, 95, 107, 109, 125, 133, 138, 140, 233, 253
Natural Bridges State Beach, 267
Nemalion andersonii. See hairy seaweed (*Cumagloia andersonii*)
Nemalion helminthoides (rubber threads), 238
Nemertea (phylum), 39–43
Nemertes gracilis. See green ribbon worm (*Emplectonema gracile*)
Neoagardhiella baileyi. See succulent seaweed (*Sarcodiotheca gaudichaudii*)
Neoagardhiella gaudichaudii. See succulent seaweed (*Sarcodiotheca gaudichaudii*)
Neomolgus littoralis (red velvet mite), 187
Neoptilota densa (half-fringed sea fern), 248
Neorhodomela larix (black pine), 250
Neotrypaea californiensis (bay ghost shrimp), 171
Nephtys sp. (shimmy worm), 45
Nereis vexillosa (pile worm), 45
Nereocystis luetkeana (bull kelp), 233
Netastoma rostrata. See beaked piddock (*Netastoma rostratum*)
Netastoma rostratum (beaked piddock), 154
Nettastomella rostrata. See beaked piddock (*Netastoma rostratum*)
netted blade (*Dictyoneurum californicum*), 233
Neverita reclusiana (southern moonsnail), 95
Nitidella carinata (carinate dovesnail *Alia carinata*), 102
Nitidiscala tincta (tinted wentletrap), 87
Nitophyllum ruprechtianum. See fringed hidden rib (*Cryptopleura ruprechtiana*)
noble Pacific doris (sea lemon *Anisodoris nobilis*), 118
nori (laver *Porphyra* sp.), 238
Norrisia norrisi (smooth brown turban), 82
Norris shell (smooth brown turban *Norrisia norrisi*), 82
norrissnail (smooth brown turban *Norrisia norrisi*), 82
Norris' top shell (smooth brown turban *Norrisia norrisi*), 82
Norris' top snail (smooth brown turban *Norrisia norrisi*), 82
north Atlantic kellia (kellyclam *Kellia suborbicularis*), 136
northern clingfish (*Gobiesox maeandricus*), 216
northern horsemussel (*Modiolus modiolus*), 128
northern kelp crab (shield-backed kelp crab *Pugettia producta*), 179
northern red anemone (painted anemone *Urticina crassicornis*), 33
northern ribbed mussel (ribbed mussel *Geukensia demissa*), 127
northern white slippersnail (*Crepidula nummaria*), 91
northwest ugly clam (rock entodesma *Entodesma navicula*), 157
Notoacmaea insessa. See seaweed limpet (*Discurria insessa*)
Notoacmea persona (mask limpet *Tectura persona*), 76
Notoacmea scutum (plate limpet *Tectura scutum*), 76
Notocomplana acticola (tapered flatworm), 38
Notoplana acticola. See tapered flatworm (*Notocomplana acticola*)
Nucella canaliculata (channelled dogwinkle), 99
Nucella emarginata (striped dogwinkle), 100
Nucella lamellosa (frilled dogwinkle), 99
nudibranch allies, 124
nudibranchs, 112–24
Nuttallia nuttallii (California mahogany-clam), 144
Nuttallina californica (California spiny chiton), 64
Nuttallina fluxa. See southern spiny chiton (*Nuttallina scabra*)
Nuttallina scabra (southern spiny chiton), 65
Nuttall's chiton (California spiny chiton *Nuttallina californica*), 64
Nuttall's cockle (*Clinocardium nuttallii*), 137
Nuttall's hornmouth (*Ceratostoma nuttalli*), 97
Nuttall's mahogany clam (California mahogany-clam *Nuttallia nuttallii*), 144

Nuttall's thorn purpura (Nuttall's hornmouth *Ceratostoma nuttalli*), 97
oar weed (*Laminaria sinclairii*), 229
Obelia sp. (wine-glass hydroid), 22
Ocenebra circumtexta (circled rocksnail), 100
Ocenebra lurida. *See* lurid rocksnail (*Ocinebrina lurida*)
Ocenebra poulsoni. *See* Poulson's rock snail (*Roperia poulsoni*)
ocher star (purple star *Pisaster ochraceus*), 197
Ocinebrina lurida(lurid rocksnail), 101
Ocrhestoidea californiana. *See* California beach hopper (*Megalorchestia californiana*)
octopods, 159
Octopus bimaculoides (two-spotted octopus), 159
Octopus rubescens (red octopus). *See under* two-spotted octopus (*Octopus bimaculoides*)
Oedignathus inermis (granular claw crab), 175
old growth kelp (*Pterygophora californica*), 230
Oligocottus maculosus (tidepool sculpin), 216
olive green isopod (Vosnesensky's isopod *Idotea wosnesenskii*), 167
Olivella baetica (baetic olive), 107
Olivella biplicata (purple olive), 107
Olivella pycna (zigzag olive), 108
olives, 107–8
Olympia oyster (*Ostrea conchaphila*), 129
Onchidoris bilamellata (barnacle-eating dorid), 115
Onchidoris fusca. *See* barnacle-eating dorid (*Onchidoris bilamellata*)
onyx slipper-shell (onyx slippersnail *Crepidula onyx*), 90
onyx slippersnail (*Crepidula onyx*), 90
onyx slipper snail (onyx slippersnail *Crepidula onyx*), 90
opalescent nudibranch (*Hermissenda crassicornis*), 124
Opalia borealis (boreal wentletrap), 88
Opalia chacei (boreal wentletrap *Opalia borealis*), 88

Opalia crenimarginata. *See* scallop-edged wentletrap (*Opalia funiculata*)
Opalia funiculata (scallop-edged wentletrap), 88
Opalia insculpta. *See* scallop-edged wentletrap (*Opalia funiculata*)
Opalia wroblewskyi. *See* boreal wentletrap (*Opalia borealis*)
Ophioderma panamense (Panamanian brittle star), 202
Ophioderma panamensis (Panamanian brittle star *Ophioderma panamense*), 202
Ophiodermella incisa (gray snakeskin-snail *Ophiodermella inermis*), 110
Ophiodermella inermis (gray snakeskin-snail), 110
Ophiodermella ophioderma (gray snakeskin-snail *Ophiodermella inermis*), 110
Ophiodromus pugettensis (bat star worm), 44
Ophionereis annulata (banded brittle star), 200
Ophiopholis aculeata (daisy brittle star), 199
Ophioplocus esmarki (smooth brittle star), 201
Ophioptereis papillosa (flat-spined brittle star *Ophiopteris papillosa*), 201
Ophiopteris papillosa (flat-spined brittle star), 201
Ophiopterus papillosa (flat-spined brittle star *Ophiopteris papillosa*), 201
Ophiothrix spiculata (western spiny brittle star), 192, 200
Ophiuroidea (class), 198–202
Ophlitaspongia pennata (red encrusting sponge), 18
Opisthobranchia (subclass), 112–24
orange chain tunicate (*Botrylloides diegensis*), 213
Orange County, 282–84
orange cup coral (*Balanophyllia elegans*), 28
orange nemertean (orange ribbon worm *Tubulanus polymorphus*), 40
orange-red coral (orange cup coral *Balanophyllia elegans*), 28
orange ribbon worm (*Tubulanus polymorphus*), 40
orange-spotted nudibranch (*Triopha catalinae*), 115
Oregon rock crab (pygmy rock crab *Cancer oregonensis*), 182
Oregonia gracilis (graceful

decorator crab), 178
Oregon rock crab (pygmy rock crab *Cancer oregonensis*), 182
ornamental blenny (mosshead warbonnet *Chirolophis nugator*), 218
Ostrea conchaphila (Olympia oyster), 129
Ostrea lurida (Olympia oyster *Ostrea conchaphila*), 129
ostrich-plume hydroid (*Aglaophenia* sp.), 23
otter-shell clam (Pacific gaper *Tresus nuttallii*), 138
oval flatworm (*Pseudoalloioplana californica*), 38
owl limpet (giant owl limpet *Lottia gigantea*), 79

Pachycheles rudis (thick-clawed porcelain crab), 177
Pachygrapsus crassipes (striped shore crab), 185
Pacific black-bristled honeycomb worm (sand-castle worm *Phragmatopoma californica*), 50
Pacific blood star (*Henricia leviuscula*), 195
Pacific blue mussel (*Mytilus trossulus*), 126
Pacific calico scallop (*Argopecten ventricosus*), 132
Pacific crab (Dungeness crab *Cancer magister*), 183
Pacific eggcockle (*Laevicardium substriatum*), 137
Pacific gaper (*Tresus nuttallii*), 138
Pacific geoduck (*Panope abrupta*), 153
Pacific half-slipper (Pacific half-slippersnail *Crepipatella dorsata*), 90
Pacific half-slippersnail (*Crepipatella dorsata*), 90
Pacific henricia (Pacific blood star *Henricia leviuscula*), 195
Pacific jewel box (Pacific left-handed jewelbox *Pseudochama exogyra*), 135
Pacific left-handed jewelbox (*Pseudochama exogyra*), 135
Pacific littleneck (*Protothaca staminea*), 150
Pacific lugworm (*Abarenicola pacifica*), 49
Pacific mole crab (*Emerita analoga*), 178
Pacific oyster (*Crassostrea gigas*), 130
Pacific razor clam (*Siliqua patula*), 140

Pacific rock crab (spot-bellied rock crab *Cancer antennarius*), 183
Pacific rockweed (*Fucus gardneri*), 235
Pacific samphire (common pickleweed *Salicornia virginica*), 251, 252
Pacific sand crab (Pacific mole crab *Emerita analoga*), 178
Pacific sand dollar (eccentric sand dollar *Dendraster excentricus*), 204
Pacific sea star (purple star *Pisaster ochraceus*), 197
Pacific shipworm (feathery shipworm *Bankia setacea*), 157
Pacific white venus (white venus *Amiantis callosa*), 146
paddle ascidian (stalked compound tunicate *Distaplia smithi*), 212
Pagurus granosimanus (grainy-hand hermit), 172
Pagurus hirsutiusculus (hairy hermit), 173
Pagurus samuelis (blueband hermit), 173
painted anemone (*Urticina crassicornis*), 33
painted brittlestar (daisy brittle star *Ophiopholis aculeata*), 198
painted nudibranch (*Dirona picta*), 122
painted spindle (*Fusinus luteopictus*), 106
painted spindle shell (painted spindle *Fusinus luteopictus*), 106
painted wentletrap (tinted wentletrap *Nitidiscala tincta*), 87
Panama brittle star (Panamanian brittle star *Ophioderma panamense*), 202
Panamanian brittle star (*Ophioderma panamense*), 202
Panamanian serpent star (Panamanian brittle star *Ophioderma panamense*), 202
Pandora punctata (punctate pandora), 158
Panope abrupta (Pacific geoduck), 153
Panopea generosa (Pacific geoduck *Panope abrupta*), 153
Panulirus interruptus (California spiny lobster), 171
papillate seaweed/sea tar (*Mastocarpus papillatus*), 246
papillose aeolid (shag-rug nudibranch *Aeolidia papillosa*), 123

papillose crab (granular claw crab *Oedignathus inermis*), 175
paralytic shellfish poisoning (PSP), 13, 130, 140, 149
Paranemertes peregrina (purple-backed ribbon worm), 43
Parapholas californica (scale-sided piddock), 155
Parastichopus californicus (California sea cucumber), 205
Parastichopus parvimensis (warty sea cucumber), 206
Pareurythoe californica (California fireworm), 46
Patiria miniata. See bat star (*Asterina miniata*)
Patrick's Point State Park, 258
pea crab (gaper pea crab *Pinnixa littoralis*), 186
peanut worm (Agassiz's peanut worm *Phascolosoma agassizii*), 55
peanut worms, 54
pea pod borer (California date-mussel *Adula californiensis*), 128
pearly jingle (*Anomia peruviana*), 133
pearly monia (green false-jingle *Pododesmus macroschisma*), 134
Pecten diegensis. See San Diego scallop (*Euvola diegensis*)
Pecten monotimeris. See kelp scallop (*Leptopecten latiauratus*)
pecten (Pacific calico scallop *Argopecten ventricosus*), 132
Pectinaria belgica. See California ice cream cone worm (*Pectinaria californiensis*)
Pectinaria californiensis (California ice cream cone worm), 51
Pectinaria granulata (trumpet worm). See under California ice cream cone worm (*Pectinaria californiensis*)
pectinate lepidozona (*Lepidozona pectinulata*), 63
Pelagia colorata (purple-striped jelly), 25
Pelagia noctiluca. See under purple-striped jelly
pelagic goose barnacle (*Lepas anatifera*), 165
Pelia tumida (dwarf teardrop crab), 181
Pelvetia fastigiata (spindle-shaped rockweed), 236
Pelvetiopsis limitata (little rockweed), 235

Penitella penita (flat-tip piddock), 155
peppered spaghetti (*Gracilaria pacifica*), 244
perennial saltwort (common pickleweed *Salicornia virginica*), 251, 252
periwinkles, 85
Peruvian jingle shell (pearly jingle *Anomia peruviana*), 133
Petrocelis franciscana. See papillate seaweed/sea tar (*Mastocarpus papillatus*)
Petrocelis franciscana. See flat-tipped wire alga (*Mastocarpus jardinii*)
Petrocelis middendorffii. See papillate seaweed/sea tar (*Mastocarpus papillatus*)
Petrolisthes cabrilloi (Cabrillo's porcelain crab), 176
Petrolisthes cinctipes (flat porcelain crab), 176
Petrolisthes eriomerus (flattop crab), 161, 175
Petrolisthes manimaculis (chocolate porcelain crab), 177
Petrospongium rugosum (convoluted sea fungus), 227
Phaeophyta (phylum), 222, 226–37
phantom shrimp (skeleton shrimp *Caprella* sp.), 168
Phascolosoma agassizii (Agassiz's peanut worm), 55
Phialidium. See clam hydroid (*Clytia bakeri*)
Phidiana crassicornis (opalescent nudibranch *Hermissenda crassicornis*), 124
Pholadidea penita. See flat-tip piddock (*Penitella penita*)
Pholis laeta (crescent gunnel), 219
Pholis ornata (saddleback gunnel), 220
Phragmatopoma californica (sand-castle worm), 50
Phyllospadix sp. (surf-grass), 253
pickleweed (*Salicornia virginica*), 252
Pikea woodii. See bleached brunette (*Cryptosiphonia woodii*)
pile worm (*Nereis vexillosa*), 45
Pilsbry's piddock (rough piddock *Zirfaea pilsbryi*), 156
Pilumnus spinohirsutus (retiring southerner), 185
pink anemone (strawberry anemone (*Corynactis californica*), 35
pink barnacle (white-ribbed red

barnacle *Megabalanus californicus*), 164
ink-fronted ribbon worm (*Amphiporus imparispinosus*), 42
ink-tipped green anemone (aggregating anemone *Anthopleura elegantissima*), 29
innixa littoralis (gaper pea crab), 186
isaster brevispinus (giant pink star), 196
isaster giganteus (giant spined star), 196
isaster ochraceus (purple star), 197
ismo clam (*Tivela stultorum*), 57, 151
*ismo State Beach, 273
lagioctenium circularis (Pacific calico scallop *Argopecten ventricosus*), 132
lanocera californica. See oval flatworm (*Pseudoalloioplana californica*)
late limpet (*Tectura scutum*), 76
latform mussel (branch-ribbed mussel *Septifer bifurcatus*), 127
latyhelminthes (phylum), 37–39
latyodon cancellatus (boring softshell-clam), 152
leurobrachia bachei (sea gooseberry), 27
locamia karykina (mucus sponge), 16, 19
lumose anemone (*Metridium senile*), 34
odarke pugettensis. See bat star worm (*Ophiodromus pugettensis*)
ododesmus cepio. See (green false-jingle *Pododesmus macroschisma*)
ododesmus macroschisma (green false-jingle), 134
*oint Dume State Beach, 278–79
*oint Fermin Marine Life Refuge, 281
*oint Lobos State Reserve, 271
*oint Pinos, 269
*oint Reyes National Seashore, 263–64
*oint St. George, 257
olinices lewisii. See Lewis's moonsnail (*Euspira lewisii*)
olinices reclusianus (southern moonsnail *Neverita reclusiana*), 95
olinices recluzianus (southern moonsnail *Neverita reclusiana*), 95

Pollicipes polymerus (goose barnacle), 165
Polynoe vittata. See red-banded commensal scaleworm (*Arctonoe vittata*)
Polyplacophora (class), 58–70
Polyporolithon parcum. See stalked coralline disk (*Clathromorphum parcum*)
Polysiphonia bipinnata (black tassel *Pterosiphonia bipinnata*), 250
popping wrack (Pacific rockweed *Fucus gardneri*), 235
porcelain crab (Cabrillo's porcelain crab *Petrolisthes cabrilloi*), 176
porcelain crab (chocolate porcelain crab *Petrolisthes maniculis*), 177
porcelain crab (flattop crab *Petrolisthes eriomerus*), 175
Porifera (phylum), 17–19
Porphyra sp. (laver), 238
Postelsia palmaeformis (sea palm), 232
Poulson's dwarf triton (Poulson's rock snail *Roperia poulsoni*), 101
Poulson's rock shell (Poulson's rock snail *Roperia poulsoni*), 101
Poulson's rock snail (*Roperia poulsoni*), 101
powder puff anemone (plumose anemone *Metridium senile*), 34
Prionitis lyallii (broad iodine seaweed), 243
proboscis worm (*Glycera* sp.), 44
proliferating anemone (*Epiactis prolifera*), 32
Protothaca laciniata (rough-sided littleneck), 149
Protothaca staminea (Pacific littleneck), 150
Pseudoalloioplana californica (oval flatworm), 38
Pseudochama exogyra (Pacific left-handed jewelbox), 135
Pteropurpura festiva (festive murex), 98
Pterorytis nuttalli (Nuttall's hornmouth *Ceratostoma nuttalli*), 97
Pterosiphonia bipinnata (black tassel), 250
Pterosiphonia robusta (black tassel *Pterosiphonia bipinnata*), 250
Pterygophora californica (old growth kelp), 230
Ptilota densa. See half-fringed sea fern (*Neoptilota densa*)

Pugettia gracilis (graceful kelp crab), 179
Pugettia producta (shield-backed kelp crab), 179
Pugettia richii (cryptic kelp crab), 180
punctata thorn drupe (spotted unicorn *Acanthina punctulata*), 96
punctate pandora (*Pandora punctata*), 158
purple-backed ribbon worm *Paranemertes peregrina*, 43
purple banded jellyfish (purple-striped jelly *Pelagia colorata*), 25
purple clam (California mahogany-clam *Nuttallia nuttallii*), 144
purple dwarf olive (purple olive *Olivella biplicata*), 107
purple encrusting sponge (*Haliclona permollis*), 19
purple fan nudibranch (Spanish shawl *Flabellina iodinea*), 122
purple-hinged rock scallop (giant rock scallop *Crassadoma gigantea*), 131
purple laver (laver *Porphyra* sp.), 238
purple olive (*Olivella biplicata*), 107
purple ribbon worm (purple-backed ribbon worm *Paranemertes peregrina*), 43
purple rock crab (purple shore crab *Hemigrapsus nudus*), 186
purple sailing jellyfish (by-the-wind sailor *Velella velella*), 24
purple sea urchin (*Strongylocentrotus purpuratus*), 203
purple shore crab (*Hemigrapsus nudus*), 186
purple spined sea urchin (purple sea urchin *Strongylocentrotus purpuratus*), 203
purple sponge (purple encrusting sponge *Haliclona permollis*), 19
purple star (*Pisaster ochraceus*), 197
purple-striped jelly (*Pelagia colorata*), 25
purple whelk (frilled dogwinkle *Nucella lamellosa*), 99
Purpura foliata (leafy hornmouth *Ceratostoma foliatum*), 97
Purpura nuttalli (Nuttall's hornmouth *Ceratostoma nuttalli*), 97
Pusula californiana (California trivia), 93

Pycnopodia helianthoides (sunflower star), 197
pygmy rock crab (*Cancer oregonensis*), 182
Pyura haustor (warty sea squirt), 209

rainbow seaweed (iridescent seaweed *Mazzaella splendens*), 247
razor clam (Pacific razor clam *Siliqua patula*), 140
Recluz's moon shell (southern moonsnail *Neverita reclusiana*), 95
Recluz's moon-shell (southern moonsnail *Neverita reclusiana*), 95
red abalone (*Haliotis rufescens*), 72
red algae (phylum Rhodophyta), 222, 237–50
red and green anemone (painted anemone *Urticina crassicornis*), 33
red and white barnacle (white-ribbed red barnacle *Megabalanus californicus*), 164
red ascidian (red sea pork *Aplidium solidum*), 211
red-banded bamboo worm (*Axiothella rubrocincta*), 49
red-banded commensal scaleworm (*Arctonoe vittata*), 48
red-banded tube worm (red-banded bamboo worm *Axiothella rubrocincta*), 49
red-beaded anemone (*Urticina coriacea*), 31
red cancer crab (red rock crab *Cancer productus*), 182
red commensal scaleworm (*Arctonoe pulchra*), 47
red crab (red rock crab *Cancer productus*), 182
red encrusting sponge (*Ophlitaspongia pennata*), 18
red fringe (*Smithora naiadum*), 237
red laver (laver *Porphyra* sp.), 238
red lobster (California spiny lobster *Panulirus interruptus*), 171
red nose (Arctic hiatella *Hiatella arctica*), 154
red nudibranch (*Rostanga pulchra*), 117
red octopus (*Octopus rubescens*). *See under* two-spotted octopus (*Octopus bimaculoides*)
red ribbon worm (orange ribbon worm *Tubulanus polymorphus*), 40
red rock crab (*Cancer productus*), 182
red rock shrimp (*Lysmata californica*), 170
red sea-leaf (*Erythrophyllum delesserioides*), 243
red sea pork (*Aplidium solidum*), 211
red sea urchin (*Strongylocentrotus franciscanus*), 203
red sponge (mucus sponge *Plocamia karykina*), 19
red sponge nudibranch (red nudibranch *Rostanga pulchra*), 117
red sponge (red encrusting sponge *Ophlitaspongia pennata*), 18
red-striped acorn barnacle (*Balanus pacificus*), 164
red-striped acorn barnacle (white-ribbed red barnacle *Megabalanus californicus*), 164
red thatched barnacle (*Tetraclita rubescens*), 163
red tide (PSP), 13, 130, 140, 149
red tube worm (calcareous tube worm *Serpula columbiana*), 54
red velvet mite (*Neomolgus littoralis*), 187
Refugio State Beach, 274
regular chiton (*Lepidozona regularis*), 63
restless ribbon worm (purple-backed ribbon worm *Paranemertes peregrina*), 43
reticulate button snail (*Trimusculus reticulatus*), 124
reticulate gadinia (reticulate button snail *Trimusculus reticulatus*), 124
retiring southerner (*Pilumnus spinohirsutus*), 185
reversed chama (Pacific left-handed jewelbox *Pseudochama exogyra*), 135
Rhodomela larix. *See* black pine (*Neorhodomela larix*)
Rhodophyta (phylum), 222, 237–50
ribbed clam (rock venus *Irusella lamellifera*), 148
ribbed horse mussel (ribbed mussel *Geukensia demissa*), 127
ribbed horsemussel (ribbed mussel *Geukensia demissa*), 127
ribbed kelp (seersucker *Costaria costata*), 230
ribbed limpet (*Lottia digitalis*), 77
ribbed mussel (*Geukensia demissa*), 127
ribbed mussel (California mussel *Mytilus californianus*), 125
ribbed topsnail (blue topsnail *Calliostoma ligatum*), 80
ribbon kelp (bull kelp *Nereocystis luetkeana*), 233
ribbon kelp (winged kelp (*Alaria marginata*), 231
ribbon worms, 39–43
right-hand chama (secret jewelbox *Chama arcana*), 135
ringed brittle star (banded brittle star *Ophionereis annulata*), 200
ringed doris (ringed nudibranch *Diaulula sandiegensis*), 120
ringed nudibranch (*Diaulula sandiegensis*), 120
ring-spotted nudibranch (ringed nudibranch *Diaulula sandiegensis*), 120
Ritterella aequalisiphonis (multi-lobed tunicate), 213
rock clam (Pacific littleneck *Protothaca staminea*), 150
rock cockle (Pacific littleneck *Protothaca staminea*), 150
rock crust (encrusting coralline algae *Lithothamnion* sp), 240
rock-dwelling clam (rock entodesma *Entodesma navicula*), 157
rock-dwelling entodesma (rock entodesma *Entodesma navicula*), 157
rock-dwelling thais (striped dogwinkle *Nucella emarginata*), 100
rock entodesma (*Entodesma navicula*), 157
rock louse (western sea roach *Ligia occidentalis*), 167
rock oyster (green false-jingle *Pododesmus macroschisma*), 134
rock oyster (Pacific left-handed jewelbox *Pseudochama exogyra*), 135
rock oyster (pearly jingle *Anomia peruviana*), 133
rock oyster (secret jewelbox *Chama arcana*), 135
rock semele (*Semele rupicola*), 146
rock venus (*Irusella lamellifera*), 148
rock venus (Pacific littleneck *Protothaca staminea*), 150
rockweed isopod (Vosnesensky's isopod *Idotea wosnesenskii*), 167

rockweed (little rockweed *Pelvetiopsis limitata*), 235
rock weed (Pacific rockweed *Fucus gardneri*), 235
rock whelk (striped dogwinkle *Nucella emarginata*), 100
Roperia poulsoni (Poulson's rock snail), 101
rose nudibranch (Hopkin's rose *Hopkinsia rosacea*), 114
rose ribbon worm (*Cerebratulus montgomeryi*), 40
Rostanga pulchra (red nudibranch), 117
rostrate piddock (beaked piddock *Netastoma rostratum*), 154
rosy bryozoan (derby hat bryozoan *Eurystomella bilabiata*), 191
rosy jackknife clam (*Solen rostriformis*), 139
rosy razor clam (rosy jackknife clam *Solen rostriformis*), 139
rough anemone (giant green anemone (*Anthopleura xanthogrammica*), 30
rough cockle (rough-sided littleneck *Protothaca laciniata*), 149
rough keyhole limpet (*Diodora aspera*), 73
rough limpet (*Lottia scabra*), 78
rough-mantled sea slug (barnacle-eating dorid *Onchidoris bilamellata*), 115
rough piddock (*Zirfaea pilsbryi*), 156
rough-sided littleneck (*Protothaca laciniata*), 149
rough-skinned lugworm (*Abarenicola claparedi*). See under Pacific lugworm (*Abarenicola pacifica*)
Royal Palms State Beach, 280
rubberneck clam (Pacific gaper *Tresus nuttallii*), 138
rubber threads (*Nemalion helminthoides*), 238
ruche (fringed hidden rib *Cryptopleura ruprechtiana*), 249
rufus tipped nudibranch (*Acanthodoris nanaimoensis*), 114

saddleback gunnel (*Pholis ornata*), 220
saddled blenny (saddleback gunnel *Pholis ornata*), 220
sail jellyfish (by-the-wind sailor *Velella velella*), 24
Salicornia bigelovii (Southern pickleweed). See under common pickleweed (*Salicornia virginica*)
Salicornia europaea (European pickleweed). See under common pickleweed (*Salicornia virginica*)
Salicornia pacifica (common pickleweed *Salicornia virginica*), 251, 252
Salicornia virginica (common pickleweed), 251, 252
salted doris (white spotted sea goddess *Doriopsilla albopunctata*), 119
salted yellow doris (white spotted sea goddess *Doriopsilla albopunctata*), 119
Salt Point State Park, 254, 261–62
salt sacs (sea sacs *Halosaccion glandiforme*), 221, 247
sand-castle worm (*Phragmatopoma californica*), 50
sand dollars, 202, 204
San Diego County, 284–87
San Diego dorid (ringed nudibranch *Diaulula sandiegensis*), 120
San Diego lamellaria (*Lamellaria diegoensis*), 92
San Diego scallop (*Euvola diegensis*), 133
sand shrimp (bay ghost shrimp *Neotrypaea californiensis*), 171
sandworm (shimmy worm *Nephtys* sp), 45
San Elijo State Beach, 284–85
Sanguinolaria nuttallii. See (California mahogany-clam *Nuttallia nuttallii*)
Sanguinolaria orcutti. See (California mahogany-clam *Nuttallia nuttallii*)
San Luis Obispo County, 272–74
San Mateo County, 266–67
Santa Barbara County, 274–75
Santa Cruz County, 267–71
Sarcodiotheca gaudichaudii (succulent seaweed), 244
Sargassum muticum (sargassum), 237
sargassum (*Sargassum muticum*), 237
Saxidomus gigantea (butter clam). See under California butter clam (*Saxidomus nuttalli*)
Saxidomus nuttalli (California butter clam), 150
scaled chiton (pectinate lepidozona *Lepidozona pectinulata*), 63

scaled worm-shell (scaly tube snail (*Serpulorbis squamigerus*), 86
scaled worm snail (scaly tube snail *Serpulorbis squamigerus*), 86
scaled wormsnail (scaly tube snail *Serpulorbis squamigerus*), 86
scale-sided piddock (*Parapholas californica*), 155
scale worm (red-banded commensal scaleworm *Arctonoe vittata*), 48
scale worm (red commensal scaleworm *Arctonoe pulchra*), 47
scallop-edged wentletrap (*Opalia funiculata*), 88
scalloped wentletrap (scallop-edged wentletrap *Opalia funiculata*), 88
scaly tube snail (*Serpulorbis squamigerus*), 86
scarlet sponge (red encrusting sponge *Ophlitaspongia pennata*), 18
scavenging isopod (*Cirolana harfordi*), 166
Schizothoerus nuttalli (Pacific gaper *Tresus nuttallii*), 138
screw shell (mudflat snail *Batillaria cumingi*), 86
Scripps Shoreline and Coastal Reserve, 285
sculptured wentletrap (scallop-edged wentletrap *Opalia funiculata*), 88
Scyphozoa (class), 24–26
sea anemones, 27–35
sea asparagus (common pickleweed *Salicornia virginica*), 251, 252
sea bat (bat star *Asterina miniata*), 193
seabee worm (sand-castle worm *Phragmatopoma californica*), 50
sea blubber (lion's mane *Cyanea capillata*), 26
sea cabbage (*Hedophyllum sessile*), 229
sea cauliflower (*Leathesia difformis*), 226
sea clown triopha (orange-spotted nudibranch *Triopha catalinae*), 115
sea cockle (white venus *Amiantis callosa*), 146
sea cradles. See chitons
sea cucumbers, 204–6
sea film (papillate seaweed/sea tar *Mastocarpus papillatus*), 246

sea gooseberry (*Pleurobrachia bachei*), 27
sea hair (tubeweed *Enteromorpha* sp.), 222
sea hare (California sea hare *Aplysia californica*), 112
sea lemon (*Anisodoris nobilis*), 118
sea lettuce (*Ulva fenestrata*), 223
sea-lichen bryozoan (*Dendrobeania lichenoides*), 190
sea mat bryozoan (sea-lichen bryozoan *Dendrobeania lichenoides*), 190
sea moss (*Cladophora* sp.), 224
sea moss (nail brush seaweed *Endocladia muricata*), 242
sea mouse (copper-haired sea mouse *Aphrodita japonica*), 46
sea mussel (California mussel *Mytilus californianus*), 125
sea nettle (*Chrysaora fuscescens*), 20, 25
sea nettle (lion's mane *Cyanea capillata*), 26
sea nipples (sea sacs *Halosaccion glandiforme*), 221, 247
sea palm (*Postelsia palmaeformis*), 232
sea pearls (*Derbesia marina*), 225
sea plume (wine-glass hydroid *Obelia* sp.), 22
sea pork (California sea pork *Aplidium californicum*), 210
sea potato (sea cauliflower *Leathesia difformis*), 226
sea sacks (sea sacs *Halosaccion glandiforme*), 221, 247
sea sacs (*Halosaccion glandiforme*), 221, 247
sea slugs, 112–24
sea squirts, 207–13
sea staghorn (*Codium fragile*), 225
sea stars, 193–202
sea tar (papillate seaweed/sea tar *Mastocarpus papillatus*), 246
sea walnut comb jelly (sea gooseberry *Pleurobrachia bachei*), 27
seaweed limpet (*Discurria insessa*), 80
seaweeds, 221–50
secret jewelbox (*Chama arcana*), 135
seersucker (*Costaria costata*), 230
segmented worms, 43–54
Semele decisa (clipped semele), 145

semele-of-the-rocks (rock semele *Semele rupicola*), 146
Semele rupicola (rock semele), 146
Septifer bifurcatus (branch-ribbed mussel), 127
serpent star (daisy brittle star *Ophiopholis aculeata*), 198
serpent star (dwarf brittle star *Amphipholis squamata*), 199
serpent star (flat-spined brittle star *Ophiopteris papillosa*), 201
Serpula columbiana (calcareous tube worm), 54
Serpula vermicularis. See under calcareous tube worm (*Serpula columbiana*)
Serpulorbis squamigerus (scaly tube snail), 86
Sertularella turgida (turgid garland hydroid), 22
sewing thread (peppered spaghetti *Gracilaria pacifica*), 244
shaggy mouse nudibranch (shag-rug nudibranch *Aeolidia papillosa*), 123
shag-rug nudibranch (*Aeolidia papillosa*), 123
Shaskyus festivus. See festive murex (*Pteropurpura festiva*)
shell binder worm (coarse-tubed pink spaghetti worm *Thelepus crispus*), 51
shell-binding colonial worm (*Chone ecaudata*), 53
shell formation, 75
shield-backed kelp crab (*Pugettia producta*), 179
shield limpet (*Lottia pelta*), 75
shimmy worm (*Nephtys* sp.), 45
shore crab (green crab *Carcinus maenas*), 184
shore liparid, 217
short-spined sea star (giant pink star *Pisaster brevispinus*), 196
short-spired purple (striped dog-winkle *Nucella emarginata*), 100
shrimps, 169–71
sickle jackknife clam (*Solen sicarius*), 140
sickle razor clam (sickle jack-knife clam *Solen sicarius*), 140
Sigillinaria aequali-siphonis (multi-lobed tunicate *Ritterella aequalisiphonis*), 213
Siliqua patula (Pacific razor clam), 140
Silver Strand State Beach, 287
Sipuncula (phylum), 55
Sitka coastal shrimp (Sitka shrimp *Heptacarpus sitchensis*), 169

Sitka shrimp (*Heptacarpus sitchensis*), 169
six-armed sea star (six-rayed star *Leptasterias hexactis* species complex), 195
six-lined nemertean (six-lined ribbon worm *Carinella sexlineata*), 39
six-lined ribbon worm (*Carinella sexlineata*), 39
six-rayed star (*Leptasterias hexactis* species complex), 195
skeleton shrimp (*Caprella* sp.), 168
slaty blue chiton (regular chiton *Lepidozona regularis*), 63
slender crab (graceful kelp crab *Pugettia gracilis*), 179
small acorn barnacle (little brown barnacle *Chthamalus dalli*), 162
small brittle star (dwarf brittle star *Amphipholis squamata*), 199
smalleye coastal shrimp (small-eyed shrimp *Heptacarpus carinatus*), 169
small-eyed coastal shrimp (small-eyed shrimp *Heptacarpus carinatus*), 169
smalleyed shrimp (*Heptacarpus carinatus*), 169
small green anemone (proliferating anemone *Epiactis prolifera*), 32
small perennial kelp (*Macrocystis integrifolia*), 234
Smithora naiadum (red fringe), 237
Smith's distaplia (stalked compound tunicate *Distaplia smithi*), 212
smooth bay shrimp (*Crangon stylirostris*), 170
smooth brittle star (*Ophioplocus esmarki*), 201
smooth brown turban (*Norrisia norrisi*), 82
smooth California venus (smooth venus *Chione fluctifraga*), 148
smooth chione (smooth venus *Chione fluctifraga*), 148
smooth crangon (smooth bay shrimp *Crangon stylirostris*), 170
smooth kelly clam (kellyclam *Kellia suborbicularis*), 136
smooth lepidozona (*Lepidozona intersticta*), 62
smooth Pacific venus (smooth venus *Chione fluctifraga*), 148
smooth porcelain crab (flat porcelain crab *Petrolisthes*

cinctipes), 176
smooth red sponge (mucus sponge *Plocamia karykina*), 19
smooth turban (smooth brown turban *Norrisia norrisi*), 82
smooth turban snail (smooth brown turban *Norrisia norrisi*), 82
smooth venus (*Chione fluctifraga*), 148
snails, 80–111
snakeskin brittle star (Panamanian brittle star *Ophioderma panamense*), 202
soft-bellied crab (granular claw crab *Oedignathus inermis*), 175
soft clam (softshell-clam *Mya arenaria*), 152
softshell-clam (*Mya arenaria*), 152
Solander's trivia (*Pusula solandri*). *See under* California trivia (*Pusula californiana*)
Solaster stimpsoni (striped sun star), 194
Solen perrini. See sickle jackknife clam (*Solen sicarius*)
Solen rosaceus. See rosy jackknife clam (*Solen rostriformis*)
Solen rostriformis (rosy jackknife clam), 139
Solen sicarius (sickle jackknife clam), 140
solitary anemone (giant green anemone *Anthopleura xanthogrammica*), 30
solitary coral (orange cup coral *Balanophyllia elegans*), 28
solitary sea squirt (warty sea squirt *Pyura haustor*), 209
Sonoma Coast State Beach, 262
Sonoma County, 261–62
southern California sea cucumber (warty sea cucumber *Parastichopus parvimensis*), 206
southern jingle (pearly jingle *Anomia peruviana*), 133
southern kelp crab (globose kelp crab *Taliepus nuttalli*), 180
southern moon shell (southern moonsnail *Neverita reclusiana*), 95
southern moonsnail (*Neverita reclusiana*), 95
southern pickleweed (*S. bigelovii*). *See under* common pickleweed (*Salicornia virginica*)
southern sea palm (double pompom kelp *Eisenia arborea*), 232

southern spiny chiton (*Nuttallina scabra*), 65
Sowerby's paper bubble (green bubble *Haminoea virescens*), 111
Spanish shawl (*Flabellina iodinea*), 122
speckled limpet (mask limpet *Tectura persona*), 76
speckled scallop (Pacific calico scallop *Argopecten ventricosus*), 132
speckled sea lemon (sea lemon *Anisodoris nobilis*), 118
speckled tegula (speckled turban *Tegula gallina*), 83
speckled turban (*Tegula gallina*), 83
spindle-shaped rockweed (*Pelvetia fastigiata*), 236
spiny brittle star (western spiny brittle star *Ophiothrix spiculata*), 200
spiny Christmas tree worm (*Spirobranchus spinosus*), 52
spiny cockle (*Trachycardium quadragenarium*), 138
spiny cup-and-saucer (*Crucibulum spinosum*), 91
spiny cup and saucer shell (spiny cup-and-saucer *Crucibulum spinosum*), 91
spiny cup-and-saucer shell (spiny cup-and-saucer *Crucibulum spinosum*), 91
spiny-headed sea squirt (*Boltenia villosa*), 203
spiny-headed tunicate (spiny-headed sea squirt *Boltenia villosa*), 203
spiny pricklycockle (spiny cockle *Trachycardium quadragenarium*), 138
spiny-skinned animals, 192–206
spiny spiral-gilled tube worm (spiny Christmas tree worm *Spirobranchus spinosus*), 52
spiral tube worm (*Spirobis* sp.), 53
Spirobranchia spinosus (spiny Christmas tree worm *Spirobranchus spinosus*), 52
Spirobranchus spinosus (spiny Christmas tree worm), 52
Spirobis sp. (spiral tube worm), 53
Spisula hemphillii. See Hemphill surfclam (*Mactromeris hemphillii*)
split kelp (*Laminaria setchellii*), 228
sponges, 16–19
sponge seaweed (sea staghorn *Codium fragile*), 225

Spongomorpha coalita. See tangle weed (*Acrosiphonia coalita*)
spongy cushion (*Codium setchellii*), 226
spoonworms, 56
spot-bellied rock crab (*Cancer antennarius*), 183
spotted dirona (painted nudibranch *Dirona picta*), 122
spotted nudibranch (ringed nudibranch *Diaulula sandiegensis*), 120
spotted thorn drupe (angular unicorn *Acanthina spirata*), 95
spotted thorn drupe (spotted unicorn *Acanthina punctulata*), 96
spotted triopha (*Triopha maculata*), 116
spotted unicorn (*Acanthina punctulata*), 96
squids, 159
staghorn fringe (*Ceramium codicola*), 225
stalked compound tunicate (*Distaplia smithi*), 212
stalked coralline disk (*Clathromorphum parcum*), 240
stalked hairy sea squirt (spiny-headed sea squirt *Boltenia villosa*), 203
stalked tunicate (long-stalked sea squirt *Styela montereyensis*), 209
starfish. *See* sea stars
Steinberg's dorid (cryptic nudibranch *Doridella steinbergae*), 113
Stenoplax conspicua (conspicuous chiton), 66
Stenoplax heathiana (Heath's chiton), 67
Stenoplax sarcosa. See conspicuous chiton (*Stenoplax conspicua*)
Stichopus californicus. See California sea cucumber (*Parastichopus californicus*)
Stichopus parvimensis. See warty sea cucumber (*Parastichopus parvimensis*)
stiff-footed sea cucumber (*Eupentacta quinquesemita*), 206
Stimpson's sun star (striped sun star *Solaster stimpsoni*), 194
stone hair (*Lithothrix aspergillum*), 241
straight horsemussel (*Modiolus rectus*), 129
strawberry anemone (*Corynactis californica*), 35

striped anemone (*Haliplanella luciae*), 32
striped dogwinkle (*Nucella emarginata*), 100
striped rock shrimp (red rock shrimp *Lysmata californica*), 170
striped shore crab (*Pachygrapsus crassipes*), 185
striped sun star (*Solaster stimpsoni*), 194
Strongylocentrotus franciscanus (red sea urchin), 203
Strongylocentrotus purpuratus (purple sea urchin), 203
stubby rose anemone (red-beaded anemone *Urticina coriacea*), 31
Styela montereyensis (long-stalked sea squirt), 209
suborbicular kellyclam (kellyclam *Kellia suborbicularis*), 136
succulent seaweed (*Sarcodiotheca gaudichaudii*), 244
sulfur sponge (yellow boring sponge (*Cliona celata*), 17
summer clam (Pacific gaper *Tresus nuttallii*), 138
sun anemone (plumose anemone *Metridium senile*), 34
sunflower star (*Pycnopodia helianthoides*), 197
sunset clam (California sunset clam *Gari californica*), 145
sun star (striped sun star *Solaster stimpsoni*), 194
surf anemone (aggregating anemone *Anthopleura elegantissima*), 29
surf-grass (*Phyllospadix* sp.), 253
surf mussel (California mussel *Mytilus californianus*), 125
swell shark (egg case) (*Cephaloscyllium ventriosum*), 215
swellshark (swell shark *Cephaloscyllium ventriosum*), 215
swimming isopod (scavenging isopod *Cirolana harfordi*), 166

Tagelus californianus (California jackknife clam), 141
Taliepus nuttalli (globose kelp crab), 180
tall coralline alga (tidepool coralline alga *Corallina officinalis* var. *chilensis*), 241
tangle weed (*Acrosiphonia coalita*), 224

tan peanut worm (bushy-headed peanut worm *Themiste pyroides*), 55
tapered flatworm (*Notocomplana acticola*), 38
Tapes japonica. See Japanese littleneck (*Venerupis philippinarum*)
Tapes philippinarum. See Japanese littleneck (*Venerupis philippinarum*)
tar spot (papillate seaweed/sea tar *Mastocarpus papillatus*), 246
Tealia coriacea. See red-beaded anemone (*Urticina coriacea*)
Tealia crassicornis. See painted anemone (*Urticina crassicornis*)
Tectura persona (mask limpet), 76
Tectura scutum (plate limpet), 76
Tegula aureotincta (gilded turban), 84
Tegula brunnea (brown turban), 82
Tegula eiseni (banded turban), 83
Tegula funebralis (black turban), 81
Tegula gallina (speckled turban), 83
Tegula ligulata. See banded turban (*Tegula eiseni*)
Tegula mendella (banded turban *Tegula eiseni*), 83
Tellina bodegensis (bodega tellin), 143
Telmessus cheiragonus (helmet crab), 181
Terebratalia transversa (common lampshell), 160
teredo (feathery shipworm *Bankia setacea*), 157
Tethys californica (California sea hare *Aplysia californica*), 112
Tetraclita rubescens (red thatched barnacle), 163
Tetraclita squamosa. See red thatched barnacle (*Tetraclita rubescens*)
Thais canaliculata (channelled dogwinkle *Nucella canaliculata*), 99
Thais emarginata. See striped dogwinkle (*Nucella emarginata*)
Thais lamellosa (frilled dogwinkle *Nucella lamellosa*), 99
thatched barnacle (red thatched barnacle *Tetraclita rubescens*), 163

Thelepus crispus (coarse-tubed pink spaghetti worm), 51
Thelepus setosus (fine-tubed pink spaghetti worm). See under coarse-tubed pink spaghetti worm (*Thelepus crispus*)
Themiste pyroides (bushy-headed peanut worm), 55
thick amphiporus (two-spotted ribbon worm *Amphiporus bimaculatus*), 41
thick-clawed porcelain crab (*Pachycheles rudis*), 177
thickclaw porcelain crab (thickclawed porcelain crab *Pachycheles rudis*), 177
thick-horned nudibranch (opalescent nudibranch *Hermissenda crassicornis*), 124
thick-petaled rose anemone (painted anemone *Urticina crassicornis*), 33
thin amphiporus (pink-fronted ribbon worm *Amphiporus imparispinosus*), 42
Thoracophelia. See bloodworm (*Euzonus* sp.)
thorn snail (angular unicorn *Acanthina spirata*), 95
three lined aeolid (*Flabellina trilineata*), 123
three-lined nudibranch (three lined aeolid (*Flabellina trilineata*), 123
tidepool coralline alga (*Corallina officinalis* var. *chilensis*), 241
tide pool coral (tidepool coralline alga *Corallina officinalis* var. *chilensis*), 241
tidepool sculpin (*Oligocottus maculosus*), 216
tidepool snailfish (*Liparis florae*), 217
tides, 8–9
tinted wentletrap (*Nitidiscala tincta*), 87
tiny pink clam (Baltic macoma *Macoma balthica*), 141
tiny tube worm (spiral tube worm *Spirorbis* sp.), 53
Tivela scarificata. See Pismo clam (*Tivela stultorum*)
Tivela stultorum (Pismo clam), 57, 151
Tomales Bay State Park, 263
Tonicella lineata (lined chiton), 58
Tonicella lokii (Loki's chiton), 59
Tonicella undocaerulea (blueline chiton), 59

Trachycardium quadragenarium (spiny cockle), 138
trellised chiton (pectinate lepidozona *Lepidozona pectinulata*), 63
Tremulla difformis (sea cauliflower *Leathesia difformis*), 226
Tresus nuttallii (Pacific gaper), 138
Trimusculus reticulatus (reticulate button snail), 124
Trinidad State Beach, 259
Triopha carpenteri (orange-spotted nudibranch *Triopha catalinae*), 115
Triopha catalinae (orange-spotted nudibranch), 115
Triopha grandis. See spotted triopha (*Triopha maculata*)
Triopha maculata (spotted triopha), 116
Trivia californiana (California trivia *Pusula californiana*), 93
troglodyte chiton (southern spiny chiton *Nuttallina scabra*), 65
trumpet worm (*Pectinaria granulata*). See under California ice cream cone worm (*Pectinaria californiensis*)
tubeweed (*Enteromorpha* sp.), 222
Tubulanus polymorphus (orange ribbon worm), 40
Tubulanus sexlineatus. See six-lined ribbon worm (*Carinella sexlineata*)
tunicates, 207–13
turgid garland hydroid (*Sertularella turgida*), 22
Turkish towel (*Chondracanthus corymbiferus*), 245
Turkish washcloth (papillate seaweed/sea tar *Mastocarpus papillatus*), 246
turtle crab (umbrella crab *Cryptolithodes sitchensis*), 174
twenty-rayed sea star (sunflower star *Pycnopodia helianthoides*), 197
two-spot keyhole limpet (*Fissurellidea bimaculata*), 74
two spot octopus (two-spotted octopus *Octopus bimaculoides*), 159
two-spotted keyhole limpet (two-spot keyhole limpet *Fissurellidea bimaculata*), 74
two-spotted octopus (*Octopus bimaculoides*), 159
two-spotted ribbon worm (*Amphiporus bimaculatus*), 41
Typhlogobius californiensis (blind goby), 220

Uca crenulata (Mexican fiddler), 187
Ulva fasciata var. *taeniata*. See corkscrew sea lettuce (*Ulva taeniata*)
Ulva fenestrata (sea lettuce), 223
Ulva sp. See tubeweed (*Enteromorpha* sp.)
Ulva taeniata (corkscrew sea lettuce), 223
umbrella crab (*Cryptolithodes sitchensis*), 174
unicorns, 95–96
unstable limpet (*Lottia instabilis*), 79
Upper Newport Bay Ecological Reserve, 282–83
urchins, 202–3
Urechis caupo (fat innkeeper worm), 56
Urochordata (phylum), 207–13
Urosalpinx cinerea (Atlantic oyster drill), 102
Urosalpinx lurida. See lurid rocksnail (*Ocinebrina lurida*)
Urticina coriacea (red-beaded anemone), 31
Urticina crassicornis (painted anemone), 33

variegate amphissa (variegated amphissa *Amphissa versicolor*), 103
variegated amphissa (*Amphissa versicolor*), 103
variegated dove shell (variegated amphissa *Amphissa versicolor*), 103
Velella lata. See by-the-wind sailor (*Velella velella*)
Velella velella (by-the-wind sailor), 24
veligers, 157
velvet ribbon worm (*Lineus pictifrons*), 41
velvety red sponge (red encrusting sponge *Ophlitaspongia pennata*), 18
Venerupis lamellifera. See rock venus (*Irusella lamellifera*)
Venerupis philippinarum (Japanese littleneck), 149
Ventura County, 276
Venus gibbosula. See smooth venus (*Chione fluctifraga*)
Venus neglecta. See wavy venus (*Chione undatella*)
venus's girdle (feather boa kelp *Egregia menziesii*), 231
Venus succincta. See banded venus (*Chione californiensis*)
violet volcano sponge (purple encrusting sponge *Haliclona permollis*), 19
Virginia oyster (Atlantic oyster *Crassostrea virginica*), 130
volcano barnacle (red thatched barnacle *Tetraclita rubescens*), 163
volcano keyhole limpet (*Fissurella volcano*), 73
volcano limpet (volcano keyhole limpet (*Fissurella volcano*), 73
Volsella flabellatus. See straight horsemussel (*Modiolus rectus*)
Volsella modiolus (northern horsemussel *Modiolus modiolus*), 128
Volsella recta. See straight horsemussel (*Modiolus rectus*)
Volvarina taeniolata (banded California marginella), 109
Vosnesensky's isopod (*Idotea wosnesenskii*), 167

wandering nemertean (purple-backed ribbon worm *Paranemertes peregrina*), 43
wandering ribbon worm (purple-backed ribbon worm *Paranemertes peregrina*), 43
warty sea cucumber (*Parastichopus parvimensis*), 206
warty sea squirt (*Pyura haustor*), 209
warty tunicate (warty sea squirt *Pyura haustor*), 209
Washington clam (California butter clam *Saxidomus nuttalli*), 150
Washington hoof shell (flat hoofsnail *Hipponix cranioides*), 89
water jelly (*Aequorea* sp.), 23
water jellyfish (water jelly *Aequorea* sp.), 23
wavy chione (wavy venus *Chione undatella*), 147
wavy top shell (wavy turban *Megastraea undosa*), 81
wavy top snail (wavy turban *Megastraea undosa*), 81
wavy top turban (wavy turban *Megastraea undosa*), 81
wavy turban (*Megastraea undosa*), 81
wavy turban snail (wavy turban *Megastraea undosa*), 81
wavy venus (*Chione undatella*), 147
webbed sea star (bat star *Asterina miniata*), 193
wedge clam (*Donax californicus*). See under bean clam (*Donax gouldii*)
wentletraps, 87–88
western banded tegula (banded turban *Tegula eiseni*), 83

western distaplia (mushroom tunicate *Distaplia occidentalis*), 212
western fat dog nassa (*Nassarius perpinguis*), 105
western fat nassa (western fat dog nassa *Nassarius perpinguis*), 105
western lean nassa (*Nassarius mendicus*), 105
western sea roach (*Ligia occidentalis*), 167
western spiny brittle star (*Ophiothrix spiculata*), 192, 200
Westport-Union Landing State Beach, 259–60
wheat shells (banded California marginella *Volvarina taeniolata*), 109
whelk, 99–100, 104
white amiantis (white venus *Amiantis callosa*), 146
white bubble (*Haminoea vesicula*), 111
white bubble shell (white bubble *Haminoea vesicula*), 111
white bubble snail (white bubble *Haminoea vesicula*), 111
whitecap limpet (*Acmaea mitra*), 75
white dendronotid (*Dendronotus albus*), 120
white dendronotus (white dendronotid *Dendronotus albus*), 120
white hoofsnail (flat hoofsnail *Hipponix cranioides*), 89
white intertidal ribbon worm (*Amphiphorus formidabilis*). *See under* pink-fronted ribbon worm (*Amphiporus imparispinosus*)
white-lined dirona (frosted nudibranch *Dirona albolineata*), 121
white-plumed anemone (plumose anemone *Metridium senile*), 34
White Point/Royal Palms County Beach, 280
white-ribbed red barnacle (*Megabalanus californicus*), 164
white sand clam (white-sand macoma *Macoma secta*), 143
white-sand macoma (*Macoma secta*), 143
white sea cucumber (stiff-footed sea cucumber *Eupentacta quinquesemita*), 206
white sea gherkin (stiff-footed sea cucumber *Eupentacta quinquesemita*), 206
white sea jelly (moon jelly (*Aurelia labiata*), 26
white slipper shell (northern white slippersnail *Crepidula nummaria*, 91
white slipper snail (northern white slippersnail *Crepidula nummaria*), 91
white-speckled nudibranch (white spotted sea goddess *Doriopsilla albopunctata*), 119
white spotted porostome (white spotted sea goddess *Doriopsilla albopunctata*), 119
white spotted sea goddess (*Doriopsilla albopunctata*), 119
white-streaked dirona (frosted nudibranch *Dirona albolineata*), 121
white venus (*Amiantis callosa*), 146
wide desmarestia (flat acid kelp *Desmarestia ligulata*), 228
wide-eared scallop (kelp scallop *Leptopecten latiauratus*), 132
wild nori (laver *Porphyra* sp.), 238
window seaweed (sea lettuce *Ulva fenestrata*), 223
wine-glass hydroid (*Obelia* sp.), 22
winged fronds (*Delesseria decipiens*), 249
winged kelp (*Alaria marginata*), 231
winged rib (winged fronds *Delesseria decipiens*), 249
wireweed (sargassum *Sargassum muticum*), 237
woody chiton (*Mopalia lignosa*), 68
worms, marine, 36–56
worn out dog whelk (eastern mud snail *Ilyanassa obsoleta*, 106
wrinkled amphissa (*Amphissa columbiana*), 103
wrinkled dove snail (wrinkled amphissa *Amphissa columbiana*), 103
wrinkled purple (frilled dogwinkle *Nucella lamellosa*), 99
wrinkled seapump (warty sea squirt *Pyura haustor*), 209
wrinkled slippershell (Pacific half-slippersnail *Crepipatella dorsata*), 90
wrinkled whelk (frilled dogwinkle *Nucella lamellosa*), 99

Xiphister atropurpureus (black prickleback), 219

yellow boring sponge (*Cliona celata*), 17
yellow lobed tunicate (*Eudistoma ritteri*), 211
yellow porostome (white spotted sea goddess *Doriopsilla albopunctata*), 119
yellow-edged cadlina (yellow-edged nudibranch *Cadlina luteomarginata*), 117
yellow-edged nudibranch (*Cadlina luteomarginata*), 117
yellow-gilled porostome (yellow-gilled sea goddess *Doriopsilla gemela*), 119
yellow-gilled sea goddess (*Doriopsilla gemela*), 119

zigzag olive (*Olivella pycna*), 108
Zirfaea pilsbryi (rough piddock), 156
Zonaria farlowii (banded tidepool fan), 227
Zonaria spadicea (chestnut cowrie), 93
Zostera sp. (eel-grass), 253